3230339074

7

D1344562

Guidance Note 7
Special Locations

WITHDRAWN

17th IET Wiring Regulations Seventeenth Edition
BS 7671:2008 Requirements for Electrical Installations
incorporating Amendment No 1:2011

Published by The Institution of Engineering and Technology, London, United Kingdom

The Institution of Engineering and Technology is registered as a Charity in England & Wales (no. 211014) and Scotland (no. SC038698).

 The Institution of Engineering and Technology is the new institution formed by the joining together of the IEE (The Institution of Electrical Engineers) and the IIE (The Institution of Incorporated Engineers). The new Institution is the inheritor of the IEE brand and all its products and services, such as this one, which we hope you will find useful.

First published 1998 (0 85296 601 6)
Second edition (incorporating Amendment No. 1 to BS 7671:2001) 2003 (0 85296 995 3)
Reprinted (incorporating Amendment No. 2 to BS 7671:2001) 2004
Third edition (incorporating BS 7671:2008) 2009 (978-0-86341-861-7)
Fourth edition (incorporating Amendment No. 1 to BS 7671:2008) 2011 (978-1-84919-283-5)
Reprinted 2012

Copies of this publication may be obtained from:
PO Box 96, Stevenage, SG1 2SD, UK
Tel: +44 (0)1438 767328
Email: sales@theiet.org
www.electrical.theiet.org/books

ISBN 978-1-84919-283-5

Typeset in the UK by Phoenix Photosetting, Chatham, Kent
Printed and bound in the UK by the MPG Books Group

Contents

Cooperating organisations

The Institution of Engineering and Technology acknowledges the invaluable contribution made by the following organisations in the preparation of this Guidance Note.

Association of Manufacturers of Domestic Appliances
S.A. MacConnacher BSc CEng MIEE MInstR

BEAMA Installation Ltd
P. Galbraith IEng MIET
Eur Ing M.H. Mullins BA CEng FIEE FIIE
P. Sayer IEng MIET GCGI

British Cables Association
C.K. Reed IEng MIET

BSI Committees
M Cotterell Co-convenor IEC TC 82
    WG3 and member of GEL 82
D Roberts BSc CEng MIMechE
    and Chairman of CPW/4

Event Supplier and Services Association
P. Yates MSc MIEE

Electrical Contractors' Association
G Digilio IEng FIEE ACIBSE MSLL
G Thomas MIET

Electrical Safety Council Electrical Contractors' Association of Scotland
R Cairney IEng MIET

GAMBICA Association Ltd
K. Morriss

Health and Safety Executive
K. Morton BSc CEng FIET

Healthcare Engineering Consultant
M Al Rufaie BSc CEng MIEE FIHEEM

Institute of Healthcare Engineering and Estate Management
Eur Ing P Harris BEng(Hons) FIHEEM MIEE MCIBSE

Institution of Engineering and Technology
P. Bicheno BSc(Hons) MIET
M. Coles BEng(Hons) MIET
G.D. Cronshaw CEng FIET (Editor)
P.E. Donnachie BSc CEng FIET
H.R. Lovegrove IEng FIET
Eur Ing L Markwell MSc BSc(Hons)
    CEng MIEE MCIBSE LCGI LCC
IM Reeve BSc(Hons) DMS CEng MIET
GG Willard DipEE CEng MIEE JP

Lighting Association
L.C. Barling

National Association of Professional Inspectors and Testers
W.R. Allen BEng(Hons) CEng MIEE
F Bertie MIET

National Caravan Council Ltd
M Perman BEng MIET

NICEIC
J.M. Maltby-Smith BSc(Hons)
    PG Dip Cert Ed IEng MIET

Safety Assessment Federation
I. Trueman CEng MIET MSOE MBES

Society of Electrical and Mechanical Engineers serving Local Government
C.J. Tanswell CEng MIET MCIBSE

Underfloor Heating Manufacturers' Association
Dr S Newberry BSc MSc PhD DIC FGS

Acknowledgements

References to British Standards, CENELEC Harmonization Documents and International Electrotechnical Commission (IEC) standards are made with the kind permission of BSI. Complete copies can be obtained by post from:

BSI Customer Services
389 Chiswick High Road
London W4 4AL
Tel: +44 (0)20 8996 9001
Fax: +44 (0)20 8996 7001
Email: orders@bsi-global.com

BSI operates an export advisory service – Technical Help to Exporters (+44 (0)20 8996 7111) – that can advise on the requirements of foreign laws and standards. The BSI also maintains stocks of international and foreign standards, with many English translations. Up-to-date information on BSI standards can be obtained from the BSI website: ww.bsi-global.com

The illustrations within this publication were provided by Rod Farquhar Design: www.farquhardesign.co.uk

Cover design and illustration were created by The Page Design: www.thepagedesign.co.uk

Copies of Health and Safety Executive documents and approved codes of practice (ACOP) can be obtained from:

HSE Books
PO Box 1999
Sudbury, Suffolk CO10 2WA
Tel: +44 (0)1787 881165
Email: hsebooks@prolog.uk.com
Web: http://books.hse.gov.uk/hse/public/home.jsf

The HSE website is www.hse.gov.uk

Preface

This Guidance Note is part of a series issued by the Institution of Engineering and Technology to explain and enlarge upon the requirements in BS 7671:2008, the 17th Edition of the IET Wiring Regulations, incorporating Amendment No 1:2011.

Note that this Guidance Note does not ensure compliance with BS 7671. It is intended to explain some of the requirements of BS 7671, but readers should always consult BS 7671 to satisfy themselves of compliance.

The scope generally follows that of BS 7671; the relevant Regulations and Appendices are noted in the margin. Some Guidance Notes also contain material not included in BS 7671:2008 but which was included in earlier editions of the Wiring Regulations. All of the Guidance Notes contain references to other relevant sources of information.

Electrical installations in the United Kingdom that comply with BS 7671 are likely to satisfy Statutory Regulations such as the Electricity at Work Regulations 1989, however this cannot be guaranteed. It is stressed that it is essential to establish which Statutory and other Regulations apply and to install accordingly. For example, an installation in premises subject to licensing may have requirements different from, or additional to, BS 7671 and these will take precedence.

Introduction

General

This Guidance Note has been revised to align with Amendment No 1:2011 to BS 7671:2008 *IET Wiring Regulations* 17th Edition. BS 7671:2008 (2011) includes additional sections on special locations that were not included in the previous edition, as follows: Medical locations (Section 710), and Operating and maintenance gangways (Section 729).

The special locations that were contained in the previous version of BS 7671 have been revised to align with the latest IEC and CENELEC standards and this Guidance Note has been revised accordingly.

One new area of development within international standards is a new Section 722 providing requirements for the supplies to electric vehicles. This proposed new section 7 would only apply to circuits intended to supply electric vehicles for charging purposes and would cover the part of the electrical installation providing the connection of an electric vehicle to the fixed installation. The draft proposals may include requirements for the type and current rating of socket-outlets, RCD protection, surge protection, measures of protection against electric shock, IP rating of equipment, impact protection against mechanical damage, isolation and switching and fixing arrangements, etc. Note: A new IET Code of Practice for electric vehicle charging equipment installation is expected to be available shortly.

This Guidance Note provides advice on the special installations and locations of Part 7 of BS 7671:2008 (2011), Requirements for Electrical Installations, *IET Wiring Regulations* 17th Edition, and the special installations and locations for which the IEC have published requirements.

It is to be noted that BS 7671 and this Guidance Note are concerned with the design, selection, erection, inspection and testing of electrical installations and that these documents may need to be supplemented by the requirements or recommendations of other British Standards.

110.1.3 Other standards of note are described in Regulation 110.1.3, including BS EN 60079 *Electrical apparatus for explosive gas atmospheres* (other than mines) and BS EN 50281 and BS EN 61241 *Electrical apparatus for use in the presence of combustible dust.*

Sect 700 The particular requirements of Part 7 of BS 7671 supplement or modify the general requirements contained in the remainder of the standard. Thus, particular protective measures may not be allowed or supplementary measures may be required. However, it is important to remember that in the absence of any comment or requirement in Part 7, the relevant requirements of the rest of the Regulations must be applied.

The particular requirements within some Part 7 sections place a prohibition on the use of certain measures of protection, e.g. obstacles and placing out of reach.

International and European Standards

Part 7 of BS 7671 technically aligns with the relevant CENELEC Harmonization Documents (HD) or draft Harmonization Documents (prHD). The preface to BS 7671 identifies the particular CENELEC Harmonization Documents current at the time of publication.

For those persons engaged in work outside the UK the Guidance Note advises whether BS 7671 is based on the European (HD) or the International (IEC) standard.

Contents

This Guidance Note discusses all the sections of Part 7 and includes chapters on special locations/installations not included in BS 7671 as follows:

▶ Chapter 13 Gardens
▶ Chapter 15 Small-scale embedded generators (SSEG)

The guidance is based on published IEC standards and draft CENELEC proposals.

Exclusions

The guide does not consider those special installations or equipment where the requirements are specified in other British Standards, such as:

BS EN 60079 ▶ Electrical apparatus for explosive gas atmospheres

BS EN 50281
BS EN 61241 ▶ Electrical apparatus for use in the presence of combustible dust

BS 5266 ▶ Emergency lighting

BS 5839 ▶ Fire detection and alarm systems in buildings.

Locations containing a bath or shower   1

1.1   Introduction

Introduction to Amendment No 1 to the 17th Edition

Sect 701   Section 701 of Amendment No 1 to BS 7671:2008 includes two notable changes.

Regulation 701.1 (Scope) now includes a fixed birthing pool.

701.411.3.3   Regulation 701.411.3.3 now states: Additional protection shall be provided for all low voltage circuits of the location, by the use of one or more RCDs having the characteristics
415.1.1   specified in Regulation 415.1.1. This is a significant change; previously Regulation 701.411.3.3 required additional protection to be provided for all circuits of the location.

The main changes introduced by the 17th Edition (BS 7671:2008) when it was published in January 2008 are described below for information.

Introduction to the 17th Edition

Sect 701   Section 701 of the 17th Edition includes a number of notable changes.

701.411.3.3   Regulation 701.411.3.3 now requires that additional protection shall be provided for all circuits of the location, by the use of one or more RCDs having the characteristics specified in Regulation 415.1.1. This is a significant change. Previously, only fixed current-using equipment (other than electric showers) located in zone 1 and current-using equipment (other than fixed current-using equipment such as a washing machine, if suitable for use in a bathroom, connected through a fused connection unit) located in zone 3 required 30 mA RCD protection. The updated Regulation 701.411.3.3 means that all circuits, including lighting, electric showers, heated towel rails, etc., will require 30 mA RCD protection. Zone 3 has now been removed.

415.1.1   Regulation 415.1.1 specifies the use of RCDs with a rated residual operating current ($I_{\Delta n}$) not exceeding 30 mA and an operating time not exceeding 40 ms at a residual current of 5 $I_{\Delta n}$.

701.512.3   Another significant change is introduced by Regulation 701.512.3. This now permits 230 V socket-outlets to be installed in a room containing a bath or shower provided that they are installed at least 3 m horizontally from the boundary of zone 1. This change removes the ambiguity that existed previously between locations containing a bath or shower and a bedroom containing a shower.

701.415.2   Regulation 701.415.2 introduces a further significant change regarding supplementary equipotential bonding. The regulation states that where the location containing a bath or shower is in a building with a protective equipotential bonding system in accordance
411.3.1.2   with Regulation 411.3.1.2, supplementary equipotential bonding may be omitted, provided that certain requirements are fulfilled.

1.2 Scope

701.1 The particular requirements of Section 701 apply to locations containing a fixed bath (bath tub, birthing pool) or shower and to the surrounding zones as described in the regulations. The requirements do not apply to emergency facilities in industrial areas or laboratories, on the presumption that they will only be used in an emergency. Where they are used with any regularity the requirements of the section would apply.

Part of the scope which stated previously: "For locations containing a bath or shower for medical treatment, special requirements may be necessary" has been deleted. The Department of Health presently have no special requirements and Section 701 should be applied. However, additional or special requirements may still be necessary for disabled persons.

1.3 The risks

The following information is provided to give a better understanding of why particular requirements are necessary for bathrooms and other wet locations.

Persons in bathrooms are particularly at risk due to:

1 lack of clothing, particularly footwear
2 presence of water reducing contact resistance
3 immersion in water, reducing total body resistance
4 ready availability of earthed metalwork
5 increased contact area.

1.3.1 Clothing

Clothing, particularly footwear such as shoes or boots, can greatly increase the total body resistance.

1.3.2 Body impedance

IEC publication DD IEC/TS 60479-1 *Effects of current on human beings and livestock* provides information on body impedance. Body impedance varies from person to person. The value of impedance depends on a number of factors, in particular on current path, touch voltage, duration of current flow, frequency, degree of moisture of the skin, surface area of contact, pressure exerted and temperature. The IEC document provides information on different total body impedances hand-to-hand for small, medium and large surface areas of contact in dry, water-wet and saltwater-wet conditions. For bathrooms and showers, Table 1.1 shows the varying total body impedance for a current path hand-to-hand, for large surface areas (10 000 mm²) in contact with dry and water-wet conditions for different touch voltages. Large surface areas of contact have the lowest impedances, with medium surface area contact (1 000 mm²) and small surface areas of contact (100 mm²) having impedances of progressively higher magnitudes.

1.3.3 Immersion

Immersion of a body in bath water produces large areas of contact of water and as Table 1.1 shows, this will reduce the total body impedance.

This reduction in body impedance coupled with a location that has earthed metalwork from pipes etc. makes a bathroom particularly hazardous and therefore require special precautions to be taken.

▼ Table 1.1 Total body impedances Z_T for a current path hand-to-hand a.c. 50/60 Hz, for large areas of contact in dry and water-wet conditions

| Touch voltage | Values for the total body impedance Z_T (Ω) that are not exceeded for: | | | | | |
| | 5% of population | | 50% of population | | 95% of population | |
(V)	Dry	Water-wet	Dry	Water-wet	Dry	Water-wet
25	1750	1175	3250	2175	6100	4100
50	1375	1100	2500	2000	4600	3675
75	1125	1025	2000	1825	3600	3275
100	990	975	1725	1675	3125	2950
125	900	900	1550	1550	2675	2675
150	850	850	1400	1400	2350	2350
175	825	825	1325	1325	2175	2175
200	800	800	1275	1275	2050	2050
225	775	775	1225	1225	1900	1900
400	700	700	950	950	1275	1275
500	625	625	850	850	1150	1150
700	575	575	775	775	1050	1050
1000	575	575	775	775	1050	1050
Asymptotic value = internal impedance	575	575	775	775	1050	1050

Notes:

1 Some measurements indicate that the total body impedance for the current path hand-to-foot is somewhat lower than for a current path hand-to-hand (10% to 30%).

2 For living persons the values of Z_T correspond to a duration of current flow of about 0.1 s. For longer durations Z_T values may decrease (about 10% to 20%) and after complete rupture of the skin Z_T approaches the internal body impedance Z_i.

3 For the standard value of the voltage 230 V (network-system 3N ~ 230/400 V) it may be assumed that the values of the total body impedance are the same as for a touch voltage of 225 V.

4 Values of Z_T are rounded to 25 W for dry and water-wet conditions.

1.4 Zones

701.32.2
701.32.3
701.32.4
Zones 0, 1 and 2 provide a very practical method of specifying requirements for protection of equipment against the ingress of water and protection against electric shock, etc. in a specific and unambiguous way. Equipment is either in a zone or outside a zone and this can be determined by measurement.

Figures 1.1a, b and g show the zones in relation to a bath tub.
Figures 1.1c, d and h show the zones in relation to a shower.
Figures 1.1e and f show the zones in relation to a shower without a basin from a fixed water outlet (shower head). Demountable shower heads are not considered and therefore do not alter the extent of the zone.

Horizontal or inclined ceilings, walls with or without windows, doors, floors and fixed partitions may be taken into account where these effectively limit the extent of locations containing a bath or shower as well as their zones.

Examples of the zones for a bath or shower are shown in Figures 1.1a–h.

Figure 1.1a
Zones for a bath in elevation view

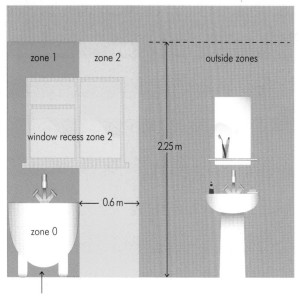

The space under the bath is:
Zone 1 if accessible without the use of a tool
Outside the zones if accessible only with the use of a tool

Figure 1.1b
Zones for a bath in plan view

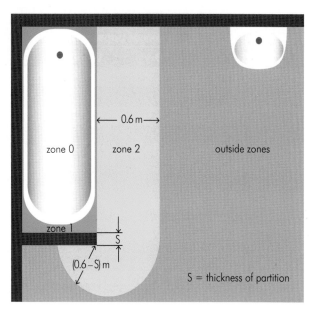

Figure 1.1c
Zones for a shower in elevation view

Figure 1.1d
Zones for a shower in plan view

Figure 1.1e
Zones for a shower, without basin

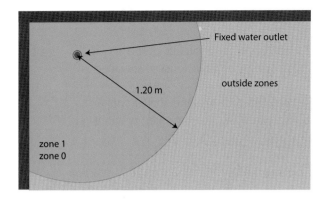

Figure 1.1f
Zones for a shower, without basin, but with permanent fixed partition

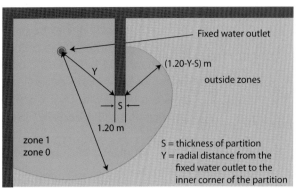

Figure 1.1g
Zones for a bath in plan view adjacent to a cupboard

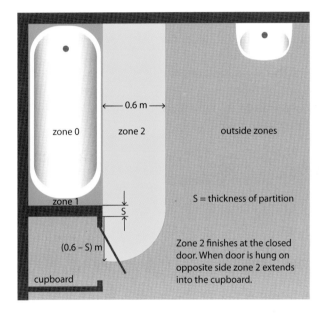

Figure 1.1h
Zones for a shower in plan view adjacent to a cupboard

1.5 Protection against electric shock

1.5.1 Locations containing a bath or shower

Section 701 includes the following requirements:

701.411.3.3 ▶ Additional protection shall be provided for all low voltage circuits of the location, by the use of one or more RCDs having the characteristics specified in Regulation 415.1.1.

701.512.3 ▶ Except for SELV socket-outlets complying with Section 414 and shaver supply
BS EN 61558-2-5 units complying with BS EN 61558-2-5, socket-outlets are prohibited within a distance of 3 m horizontally from the boundary of zone 1.

701.410.3.5 The protective measures of obstacles and placing out of reach (Section 417) are not permitted.

701.410.3.6 The protective measures of non-conducting location (Regulation 418.1) and earth-free local equipotential bonding (Regulation 418.2) are not permitted.

1.5.2 Supplementary equipotential bonding

Supplementary equipotential bonding is required to be established in all areas within
701.415.2 a room containing a bath tub or shower basin, by Regulation 701.415.2 (refer to Figure 1.2). **However, the last paragraph of the regulation allows this bonding to be omitted where the location containing a bath or shower is in a building with a**
411.3.1.2 **protective equipotential bonding system in accordance with Regulation 411.3.1.2 and provided all of the following conditions are met**: (Note: Amendment No 1 renumbers the indents below.)

  i All final circuits of the location comply with the requirements for automatic
411.3.2 disconnection according to Regulation 411.3.2
  ii All final circuits of the location have additional protection by means of an RCD in accordance with Regulation 701.411.3.3
  iii All extraneous-conductive-parts of the location are effectively connected to the protective equipotential bonding according to Regulation 411.3.1.2.

This means the designer/installer needs to establish that all extraneous-conductive-parts of the location are effectively connected to the protective equipotential bonding according to Regulation 411.3.1.2.

Note: The effectiveness of the connection of extraneous-conductive-parts in the location to the main earthing terminal may be assessed, where necessary, by
415.2.2 the application of Regulation 415.2.2.

> **415.2.2** Where doubt exists regarding the effectiveness of supplementary equipotential bonding, it shall be confirmed that the resistance R between simultaneously accessible exposed-conductive-parts and extraneous-conductive-parts fulfils the following condition:
>
> $R \leq 50$ V/I_a in a.c. systems
>
> $R \leq 120$ V/I_a in d.c. systems
>
> where:
>
> I_a is the operating current in amperes of the protective device –
>   for RCDs, $I_{\Delta n}$
>   for overcurrent devices, the current causing automatic operation in 5 s.

▼ **Figure 1.2**
Supplementary bonding
in a bathroom – metal
pipe installation

Notes:

1   The protective conductors of all power and lighting points within the zones must be
     supplementary bonded to all extraneous-conductive-parts in the zones, including metal
     waste, water and central heating pipes, and metal baths and metal shower basins.
     Supplementary bonding of circuit protective conductors and extraneous-conductive-parts
     in the zones does not require multiple connections.
2   Circuit protective conductors may be used as supplementary bonding conductors.
3   The space below the bath/shower is zone 1 if it can be accessed without the use of a
     tool, otherwise it is outside the zones.

In practice, the resistance between bonded extraneous-conductive-parts and exposed-
conductive-parts should not exceed 0.05 ohm.

If supplementary bonding is carried out, the following points are worth noting:

701.415.2 ▶ Supplementary bonding does not necessarily have to be carried out within the
               bathroom itself but may be carried out in close proximity, such as under the
               floorboards, above the ceiling, or in an adjacent airing cupboard.
           ▶ The requirement to supplementary bond to the protective conductors of circuits
               supplying both Class I and Class II equipment is necessary in case during the life
               of the installation the user changes a Class II item of equipment for Class I.

1.5.3 Supplementary bonding of plastic pipe installations

Supplementary bonding is not required to metallic parts supplied by plastic pipes such
as metal hot and cold water taps supplied from plastic pipes, or to a metal bath not
connected to extraneous-conductive-parts such as structural steelwork and where the
hot and cold water pipes and the waste are plastic (refer to Figure 1.3).

Supplementary bonding is also not required to short lengths of metal pipe that are
often installed for cosmetic reasons when the basic plumbing system is plastic.

▼ **Figure 1.3**
Supplementary bonding
in a bathroom – plastic
pipe installation

Notes:

1 The protective conductors of all power and lighting points within the zones must be supplementary bonded. The bonding connection may be to the earth terminal of a switch or accessory supplying equipment. Supplementary bonding of circuit protective conductors and extraneous-conductive-parts in the zones does not require multiple connections.

2 Circuit protective conductors may be used as supplementary bonding conductors.

3 The space below the bath/shower is zone 1 if it can be accessed without the use of a tool, otherwise it is outside the zones.

1.6   Wiring systems

701.512.2

Metal conduit, metal trunking wiring systems, and metal sheathed cables are allowed in the zones of bathrooms; however, these must be included in any supplementary bonding arrangement. They do not have to be supplying equipment within the zones, but IP ratings within zones 1 and 2 would apply.

1.7   Switchgear and controlgear

701.512.3

Switches and controls other than those which are incorporated in fixed current-using equipment suitable for use in that zone or insulating pull cords of cord-operated switches are not allowed in zones 0, 1 or 2. This means that switches are allowed on showers and fans if the IP rating of the equipment including that of the switch is appropriate for use in the particular zone.

Exceptionally, switches of SELV circuits supplied at a nominal voltage not exceeding 12 V a.c. rms or 30 V ripple-free d.c. with the safety source installed outside zones 0, 1 and 2 may be installed in zone 1, or SELV up to 50 V a.c. rms or 120 V d.c. in zone 2. Also, shaver supply units complying with BS EN 61558-2-5 may be installed in zone 2.

Except for SELV socket-outlets complying with Section 414 and shaver supply units complying with BS EN 61558-2-5, socket-outlets are prohibited within a distance of 3 m horizontally from the boundary of zone 1 (refer to Figure 1.4).

▼ **Figure 1.4**
Installation of a 230 V socket-outlet in a bathroom

The requirements for switchgear, controlgear and accessories in locations containing a bath or shower are summarised in Table 1.2.

1.7.1 Telephones

Telephones and their sockets should be installed outside zones 0, 1 and 2.

1.8 Current-using equipment

701.55 Fixed and permanently connected current-using equipment may be installed in zones 1, 2 and outside the zones, but there are specific requirements for degrees of protection (see Table 1.2 for summary of the requirements). Equipment having a rated voltage of 230 V may be installed in the above zones provided it has the appropriate IP rating and is suitable for use in the zone. This includes equipment suitable for use in the zone incorporating switches and controls.

Part 2 *Current-using equipment* is equipment that consumes current rather than simply transmits it or switches it. Examples include appliances, luminaires, fans and heaters.

Extractor fans

A suitable 230 V extractor fan may be installed in zone 1 or 2, or outside the zones.

If an extractor fan is installed in zone 1 or 2 it must be protected against the ingress of moisture to at least IPX4. An extractor fan supplied from a lighting circuit for a bathroom without a window should have its own means of isolation, as otherwise replacement or maintenance of the fan would have to be carried out in the dark. An isolation switch for a fan with an overrun facility will need to be triple-pole (switch wire, line and neutral), and must be installed outside the zones.

Note: Refer to manufacturer for overcurrent protection.

▼ **Table 1.2** Summary of requirements for equipment (current-using and accessories) in locations containing a bath or shower

Zone*	Requirements for equipment in the zones		
	Minimum degree of protection	**Current-using equipment, e.g. appliance or luminaire**	**Switchgear, controlgear and accessories**
0	IPX7	Only suitable 12 V a.c. rms or 30 V ripple-free d.c. SELV fixed and permanently connected equipment that cannot be located elsewhere, the safety source being installed outside the zones.	None allowed.
1	IPX4 (Electrical equipment exposed to water jets, e.g. for cleaning purposes, shall have a degree of protection of at least IPX5)	The following fixed and permanently connected current-using equipment if suitable for installation in zone 1 according to the manufacturer's instructions: • Whirlpool units (e.g. hot tubs and whirlpool baths) • Electric showers • Shower pumps • Equipment protected by SELV or PELV at a nominal voltage not exceeding 25 V a.c. rms or 60 V ripple-free d.c., the safety source being installed outside zones 0, 1 and 2 • Ventilation equipment • Towel rails • Water heating appliances • Luminaires	Only switches of SELV circuits supplied at a nominal voltage not exceeding 12 V a.c. rms or 30 V ripple-free d.c. shall be installed, the safety source being outside zones 0, 1 and 2.
2	IPX4 (Electrical equipment exposed to water jets, e.g. for cleaning purposes, shall have a degree of protection of at least IPX5)	Current-using equipment if suitable.	Switchgear, accessories incorporating switches or socket-outlets shall not be installed with the exception of: **i** switches and socket-outlets of SELV circuits, the safety source being installed outside zones 0, 1 and 2, and **ii** shaver supply units complying with BS EN 61558-2-5. Note: the IPX4 requirement does not apply to shaver supply units complying with BS EN 61558-2-5 installed in zone 2 and located where direct spray from showers is unlikely.
Outside zones	No additional requirement but subject to Regulation 512.2	No additional restrictions.	Except for SELV socket-outlets complying with Section 414 and shaver supply units complying with BS EN 61558-2-5, socket-outlets are prohibited within a distance of 3 m horizontally from the boundary of zone 1.

* See Figure 1.1 for zones.

1.9 Other equipment, e.g. home laundry equipment

Current-using equipment such as washing machines and tumble-driers is permitted within zone 2 if it has a minimum degree of protection of IPX4 (if subject to water jets – at least IPX5 is required) and in accordance with the manufacturer's instructions. Such equipment must be supplied by means of a permanent connection unit located outside zone 2. Beyond 3 m horizontally from the boundary of zone 1 the equipment may be supplied by means of a plug and socket.

1.10 Electric floor heating systems

701.753   For electric floor heating systems, only heating cables according to relevant product standards or thin sheet flexible heating elements according to the relevant equipment standard may be erected provided that they have either a metal sheath or metal enclosure or a fine mesh metallic grid. The fine mesh metallic grid, metal sheath or metal enclosure must be connected to the protective conductor of the supply circuit. Compliance with the latter requirement is not required if the protective measure SELV is provided for the floor heating system. For electric floor heating systems, the protective measure 'protection by electrical separation' is not permitted.

Sect 753   BS 7671:2008 includes particular requirements for floor and ceiling heating systems (Section 753), for which guidance is provided in Chapter 12.

Swimming pools and other basins

2

2.1 Introduction

Introduction to Amendment No 1 to the 17th Edition

Sect 702   Section 702 of Amendment No 1 to BS 7671:2008 includes a few small changes.

Part of the scope in previous Regulation 702.1 which stated: "Special requirements may be necessary for swimming pools for medical purposes" has been deleted.

702.410.3.4.3   The note regarding TN-C-S systems in Regulation 702.410.3.4.3 has been modified, and now refers to PME.

The main changes introduced by the 17th Edition (BS 7671:2008) when it was published in January 2008 are described below for information.

Introduction to the 17th Edition

Sect 702   Section 702 of the 17th Edition includes a number of notable changes, and it now also applies to the basins of fountains.

Zones A, B and C in the 16th Edition are replaced by zones 0, 1 and 2 (although the dimensions of the zones remain the same), except that there is no zone 2 for fountains.

A solution is included for the installation of 230 volt luminaires for swimming pools where there is no zone 2 (previously zone C). The 17th Edition states that for swimming pools where there is no zone 2, lighting equipment supplied by other than a SELV source at 12 V a.c rms or 30 V ripple-free d.c. may be installed in zone 1 on a wall or on a ceiling, provided that certain requirements are fulfilled.

Every luminaire must have an enclosure providing Class II or equivalent insulation and providing protection against mechanical impact of medium severity. This is a significant change from the 16th Edition, which only allowed SELV luminaires in zones A and B.

702.11 ## 2.2 Scope

The particular requirements of Section 702 apply to the basins of swimming pools, the basins of fountains and the basins of paddling pools. The particular requirements also apply to the surrounding zones of these basins.

In these areas, in normal use, the risk of electric shock is increased by a reduction in body resistance and contact of the body with Earth potential. Swimming pools within the scope of an equipment standard are outside the scope of the regulations. Special requirements may be necessary for swimming pools for medical purposes. Except for areas especially designed as swimming pools, the requirements do not apply to natural waters, lakes in gravel pits, coastal areas and the like.

For private garden ponds and private garden fountains see Chapter 13.

2.3 The risks

The risk of electric shock is increased in swimming pools and their surrounding zones by the reduction in body resistance (see also bathrooms, section 1.3) and by good contact with Earth arising from wet partially clothed bodies. Equipment installed close to swimming pools and fountains is required to have appropriate degrees of protection against ingress of water.

2.4 Zones

702.32 The requirements for the classification of external influences are based on the dimensions of three zones (examples are given in Figures 702.1 to 702.4 of BS 7671:2008). For a swimming pool, zones 1 and 2 may be limited by fixed partitions having a minimum height of 2.5 m (see Figure 2.1).

▼ **Figure 2.1**
Example of zone dimensions (plan) with fixed partitions of height at least 2.5 m

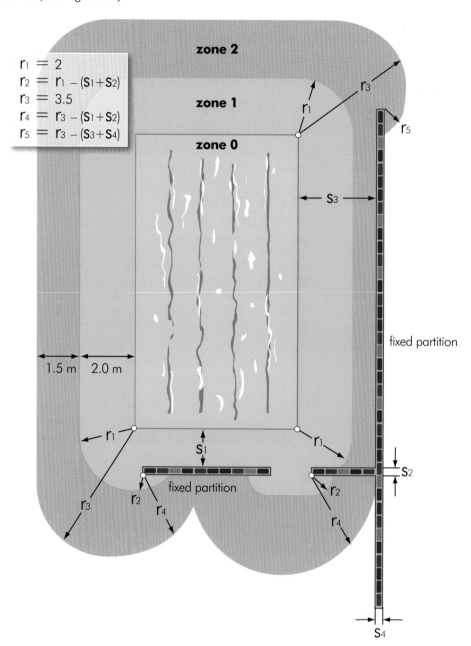

$$r_1 = 2$$
$$r_2 = r_1 - (s_1 + s_2)$$
$$r_3 = 3.5$$
$$r_4 = r_3 - (s_1 + s_2)$$
$$r_5 = r_3 - (s_3 + s_4)$$

2.4.1 Zone 0

This zone is the interior of the basin of the swimming pool or fountain including any recesses in its walls or floors, basins for foot cleaning and water jets or waterfalls and the space below them.

2.4.2 Zone 1

This zone is limited by:

▶ zone 0
▶ a vertical plane 2 m from the rim of the basin
▶ the floor or surface expected to be occupied by persons
▶ the horizontal plane 2.5 m above the floor or the surface expected to be occupied by persons.

Where the swimming pool or fountain contains divingboards, springboards, starting blocks, chutes or other components expected to be occupied by persons, zone 1 comprises the zone limited by:

▶ a vertical plane situated 1.5 m from the periphery of the divingboards, springboards, starting blocks, chutes and other components such as accessible sculptures, viewing bays and decorative basins
▶ the horizontal plane 2.5 m above the highest surface expected to be occupied by persons.

2.4.3 Zone 2

This zone is limited by:

▶ the vertical plane external to zone 1 and a parallel plane 1.5 m from the former
▶ the floor or surface expected to be occupied by persons
▶ the horizontal plane 2.5 m above the floor or surface expected to be occupied by persons.

There is no zone 2 for fountains (see Figure 2.2).

▼ **Figure 2.2**   Example of determination of the zones of a fountain

Water in zone 0
Pool, basin, waterfall and fountain space

Air Spray in zone 0
Volumes below waterjets and waterfalls to be considered as zone 0

Zone 1

2.5 Protection for safety

702.410.3.5
702.410.3.6

Not surprisingly, the protective measures of obstacles and placing out of reach (Section 417) are not permitted. Also, the protective measures of non-conducting location (Regulation 418.1) and earth-free local equipotential bonding (Regulation 418.2) are not permitted.

702.415.2

702.52

Supplementary equipotential bonding is required between all extraneous-conductive-parts and the protective conductors of all exposed-conductive-parts in zones 0, 1 and 2, irrespective of whether they are simultaneously accessible, in accordance with Regulation 415.2. In zones 0, 1 and 2, any metallic sheath or metallic covering of a wiring system (e.g. conduit), either surface-run or embedded in walls, floors etc. at a depth not exceeding 50 mm, must be connected to the supplementary equipotential bonding.

GN5

See Guidance Note 5 for test methods of determining an extraneous-conductive-part.

702.55.1

There is no particular requirement to install a metal grid in solid floors. However, where there is a metal grid, it must be connected to the local supplementary bonding. It is important to note that Section 753 'Floor and ceiling heating systems' has requirements (such as a metal grid) where electric floor warming is installed.

The requirements for electric floor warming in swimming pool locations would be similar to the requirements for locations containing a bath or shower (see section 1.10).

2.6 Luminaires

2.6.1 Luminaires for swimming pools

702.410.3.4.1

In zones 0 and 1, only protection by SELV is permitted.

In zone 0, only protection by SELV at a nominal voltage not exceeding 12 V a.c. rms or 30 V ripple-free d.c. is permitted, the source for SELV being installed outside zones 0, 1 and 2.

In zone 1, only protection by SELV at a nominal voltage not exceeding 25 V a.c. rms or 60 V ripple-free d.c. is permitted, the source for SELV being installed outside zones 0, 1 and 2.

Note: Equipment for use in the interior of basins which is only intended to be in operation when people are not inside zone 0 shall be supplied by a circuit protected by SELV, automatic disconnection of the supply (Section 411) using a 30 mA RCD, or electrical separation. See Regulation 702.410.3.4.1.

702.410.3.4.3

In zone 2, luminaires are required to be protected either by SELV, a 30 mA RCD or electrical separation. Luminaires can be protected by 30 mA RCDs, but the high protective conductor currents often found in such equipment must be carefully considered. The luminaire manufacturer's advice should be sought with the objective of determining the standing protective conductor current and the maximum protective conductor current during starting so that the luminaires do not cause unwanted tripping of the 30 mA RCD. If RCD protection is used, then the luminaires should be on more than one circuit with separate RCDs.

Where there is no zone 2, section 2.8.1 describes the requirements to be fulfilled to enable 230 V luminaires to be installed in zone 1.

2.6.2 Underwater luminaires for swimming pools

702.55.2 A luminaire for use in the water or in contact with the water must be fixed and must comply with BS EN 60598-2-18.

Underwater lighting located behind watertight portholes, and serviced from behind must comply with the appropriate part of BS EN 60598 and be installed in such a way that no intentional or unintentional conductive connection between any exposed-conductive-part of the underwater luminaires and any conductive parts of the portholes can occur.

2.7   Socket-outlets

Socket-outlets must not be installed in zones 0 or 1, except that for a swimming pool where it is not possible to install a socket-outlet outside zone 1, a socket-outlet that preferably has a non-conducting cover or coverplate, may be installed in zone 1 provided that:

▶ it is installed at least 1.25 m from the border of zone 0 and at least 0.3 m above the floor, and

702.53 ▶ the supply to the socket-outlet is protected by:
   – SELV (Section 414), at a nominal voltage not exceeding 25 V a.c. rms or 60 V ripple-free d.c., with the source for SELV being installed outside zones 0 and 1, or
   – automatic disconnection of supply (Section 411), using an RCD having the characteristics specified in Regulation 415.1.1, or
   – electrical separation (Section 413), supplying only one item of current-using equipment, the source for electrical separation being installed outside zones 0 and 1.

Socket-outlets may be installed in zone 2 provided the supply circuit is protected by:

i   SELV, with the source being installed outside zones 0, 1 and 2, or
ii automatic disconnection of supply, using an RCD having the characteristics specified in Regulation 415.1.1, or
iii electrical separation, with the source for electrical separation supplying only one socket-outlet and being installed outside zones 0, 1 and 2.

If using SELV or electrical separation to supply a socket-outlet in zone 2, the source can be installed in zone 2 provided the supply circuit is protected by an RCD having the characteristics specified in Regulation 415.1.1.

The 16th Edition used to require socket-outlets in these locations to comply with BS EN 60309-2, but this is no longer a requirement as it was changed by BS 7671:2008.

2.8 Current-using equipment of swimming pools

702.55.1   In zones 0 and 1, it is only permitted to install fixed current-using equipment specifically designed for use in a swimming pool, in accordance with the requirements of Regulations 702.55.2 and 702.55.4.

Equipment which is intended to be in operation only when people are outside zone 0 may be used in all zones provided that it is supplied by a circuit protected according to Regulation 702.410.3.4.

It is permitted to install an electric heating unit embedded in the floor, provided that it:

i   is protected by SELV (Section 414), the source of SELV being installed outside zones 0, 1 and 2. However, it is permitted to install the source of SELV in zone 2 if its supply circuit is protected by an RCD having the characteristics specified in Regulation 415.1.1, or

ii   incorporates an earthed metallic sheath connected to the supplementary equipotential bonding specified in Regulation 702.415.2 and its supply circuit is additionally protected by an RCD having the characteristics specified in Regulation 415.1.1, or

iii   is covered by an embedded earthed metallic grid connected to the supplementary equipotential bonding specified in Regulation 702.415.2 and its supply circuit is additionally protected by an RCD having the characteristics specified in Regulation 415.1.1.

Note: A socket-outlet and control device of equipment that is intended to be used in the interior of a swimming pool when not occupied should have a notice to warn the user that the equipment should only be used when the swimming pool is not occupied by persons.

2.8.1 Special requirements for the installation of electrical equipment in zone 1 of swimming pools and other basins

702.55.4   Fixed equipment designed for use in swimming pools and other basins (e.g. filtration systems, jet stream pumps) and supplied at low voltage is permitted in zone 1, subject to all the following requirements being met:

▶   The equipment must be located inside an insulating enclosure providing at least Class II or equivalent insulation and providing protection against mechanical impact of medium severity (AG2). This applies irrespective of the classification of the equipment.

▶   The equipment must only be accessible via a hatch (or a door) by means of a key or a tool. The opening of the hatch (or door) must disconnect all live conductors. The supply cable and the main disconnecting means should be installed in a way that provides protection of Class II or equivalent insulation.

▶   The supply circuit of the equipment must be protected by:
  −   SELV at a nominal voltage not exceeding 25 V a.c. rms or 60 V ripple-free d.c., the source of SELV being installed outside zones 0, 1 and 2, or
  −   an RCD having the characteristics specified in Regulation 415.1.1, or
  −   electrical separation (Section 413), the source for electrical separation supplying a single fixed item of current-using equipment and being installed outside zones 0, 1 and 2.

A solution is included for the installation of 230 volt luminaires for swimming pools where there is no zone 2.

Regulation 702.55.4 states that for swimming pools where there is no zone 2, lighting equipment supplied by other than a SELV source at 12 V a.c. rms or 30 V ripple-free d.c. may be installed in zone 1 on a wall or on a ceiling, provided that the following requirements are fulfilled:

i   the circuit is protected by automatic disconnection of the supply and additional protection is provided by an RCD having the characteristics specified in Regulation 415.1.1, and

ii the height from the floor is at least 2 m above the lower limit of zone 1.

In addition, every luminaire must have an enclosure providing Class II or equivalent insulation and providing protection against mechanical impact of medium severity.

This is a significant change from the 16th Edition, which only allowed SELV luminaires in zones A and B (i.e. zones 0 and 1).

2.9 Fire alarms and public address systems

Electrical equipment for safety systems such as fire alarms and public address may be required within the pool area. This equipment should be accessible for maintenance and preferably placed outside zones 0 and 1. Where it is not possible to locate equipment outside zone 1, it should be more than 1.25 metres outside zone 0 and as high above floor level as is practicable, in order to keep the equipment dry. The equipment will need to be of an insulated construction and have IP coding as required by the zone. Local microphones must be SELV or connected via isolating transformers, and telephones, if required, should be cordless-type within the zones with a base unit installed outside the zones.

Reference should be made to the following standards:

▶ BS 5839 – Fire detection and fire alarm systems for buildings
▶ BS EN 54 – Fire detection and fire alarm systems
▶ BS 7445/BS EN 60849 – Sound systems for emergency purposes
▶ BS 7827 – Code of practice for designing, specifying, maintaining and operating emergency sound systems at sports venues.

2.10 Fountains

2.10.1 General

The basins of fountains and their surroundings, unless persons are prevented from gaining access to them without the use of ladders or similar means, are treated as swimming pools.

702.410.3.4.2 In zones 0 and 1, one or more of the following protective measures must be used:

▶ SELV, the source for SELV being installed outside zones 0 and 1
▶ Automatic disconnection of supply using an RCD having the characteristics specified in Regulation 415.1.1
▶ Electrical separation, the separation source supplying only one item of current-using equipment and installed outside zones 0 and 1.

2.10.2 Electrical equipment of fountains

702.55.3 Electrical equipment in zones 0 or 1 must be provided with mechanical protection to medium severity (AG2), e.g. by use of mesh glass or by grids which can only be removed by the use of a tool.

Luminaires installed in zones 0 or 1 are required to be fixed and must comply with BS EN 60598-2-18.

An electric pump must comply with the requirements of BS EN 60335-2-41.

2.10.3 Additional requirements for the wiring of fountains

702.522.23 Cables supplying equipment in zone 0 should be installed outside the basin, i.e. in zone 1 or beyond where possible, and run to the equipment in the basin by the shortest practicable route.

For cables supplying equipment in zone 0, the designer should check with the supplier that the cable type is suitable for continuous immersion in water.

Cables supplying equipment in zone 1 should be selected, installed and provided with mechanical protection to medium severity (AG2) and the relevant submersion in water depth (AD8). Cable type H07RN8-F (BS 7919) is suitable up to a depth of 10 m of water. For depths greater than 10 m the cable manufacturer should be consulted.

Rooms and cabins containing sauna heaters   3

3.1   Introduction

Amendment No 1 to the 17th Edition

There are no significant changes introduced in Section 703 by Amendment No 1 to the 17th Edition.

Introduction to the 17th Edition

Sect 703   While Section 703 'Rooms and cabins containing sauna heaters' retains most of the 16th Edition requirements, there are some notable changes in the 17th Edition.

The zones have been simplified and re-designated 1, 2 and 3 in place of the previous A, B, C and D (zones C and D have been combined).

703.411.3.3   Regarding the supplies to equipment in saunas, additional protection must now be provided for all circuits of the sauna, by means of one or more RCDs having the characteristics specified in Regulation 415.1.1. However, RCD protection need not be provided for the sauna heater itself unless such protection is recommended by the manufacturer, whose advice should be sought.

3.2   Scope

703.1   The particular requirements of Section 703 apply to:

i   sauna cabins erected on site, e.g. in a location or in a room
ii   the room where the sauna heater is, or the sauna heating appliances are installed. In this case the whole room is considered as the sauna.

The requirements do not apply to prefabricated sauna cabins complying with a relevant equipment standard.

Where facilities such as showers, etc. are installed, the requirements of Section 701 will also apply.

3.3   The risks

There are two particular aspects of saunas that make them special locations:

1   Increased risk of electric shock because of extremely high humidity, lack of clothing, reduced skin resistance and large contact areas
2   Very high temperatures in certain zones.

3.3.1 Zones

703.32
Fig 703

The zones are temperature zones, dimensioned down from the ceiling, up from the floor and around the sauna heater. This allows application of the zones whatever the size of the sauna cabin.

3.4 Shock protection

703.53

Protection against electric shock is provided by not allowing any electrical equipment that is not part of the heating appliance or strictly necessary for the operation of the sauna, such as sauna thermostat, sauna cut-out and luminaires. Light switches must be placed outside the sauna room or cabin and socket-outlets should not be installed in the location containing the sauna heater. It is advisable not to install sockets near the cabin; the same criteria as for swimming pools should be adopted.

703.512.2

Only the sauna heater and equipment belonging to the sauna heater should be installed in zone 1.

703.1

A sauna is often part of a health or fitness suite and may be associated with a swimming pool, showers or bathing facilities. Such premises should be considered as a whole, and it must be borne in mind that the sauna cabin may well be located within the zones of the swimming pool.

The requirements for basic protection and fault protection in saunas are similar to those for bathrooms and swimming pools.

703.410.3.5
703.410.3.6

The protective measures of obstacles and placing out of reach (Section 417) are not permitted. Also, the protective measures of non-conducting location (Regulation 418.1) and earth-free local equipotential bonding (Regulation 418.2) are not permitted.

3.5 Wiring system

703.52

The wiring system should preferably be installed outside the zones, i.e. on the cold side of the thermal insulation. If the wiring system is installed on the warm side of the thermal insulation in zones 1 or 3, it must be heat-resisting. Any metallic sheaths or metallic conduits must not be accessible to persons using the sauna.

3.6 Heating elements

703.55

Sauna heating appliances should comply with BS EN 60335-2-53 and be installed in accordance with the manufacturer's instructions.

The heating elements incorporated in a sauna are likely to be metal sheathed. These, unless specified as having waterproof seals, may absorb moisture and cause the operation of a 30 mA RCD, if installed. RCD protection need not be provided for the sauna heater unless such protection is recommended by the manufacturer (see 3.1).

Construction and demolition site installations

4

4.1 Introduction

Introduction to Amendment No 1 to the 17th Edition

Sect 704   Section 704 of Amendment No 1 to BS 7671:2008 includes some notable changes.

704.410.3.6   Under 'Protection for safety', a new regulation has been introduced which prohibits the protective measures of non-conducting location and earth-free local equipotential
704.411.3.1   bonding. Regulation 704.411.3.1 has been modified, and now refers to PME. Also editorial correction of the last line of Regulation 704.411.3.2.1 has been made.

The main changes introduced by the 17th Edition (BS 7671:2008) when it was published in January 2008 are described below for information.

Introduction to the 17th Edition

Sect 704   While Section 704 'Construction and demolition site installations' retains the majority of the 16th Edition requirements, there are some notable changes in the 17th Edition.

The 0.2 s maximum disconnection time has been removed and replaced by clear requirements for the supply to a socket-outlet with a rated current up to and including 32 A and a circuit supplying hand-held equipment with a rated current up to and including 32 A.

704.411.3.1   For automatic disconnection of supply a TN-C-S system (but see Amendment No 1 above) is not to be used for the supply to a construction site, except for the supply to a fixed building of the construction site.

Section 704 should be read in conjunction with BS 7375 *Code of practice for distribution of electricity on construction and building sites*.

4.2 Scope

704.1   Section 704 applies to all sites of construction work including the repair or alteration of existing buildings and demolition work.

The requirements apply to fixed or movable installations. The regulations do not apply to:

▶ installations covered by the IEC 60621 series 2, where equipment of a similar nature to that used in surface mining applications is involved
▶ installations in administrative locations of construction sites (e.g. offices, cloakrooms, meeting rooms, canteens, restaurants, dormitories, toilets), where the general requirements of Parts 1 to 6 of BS 7671 apply.

4.3 The risks

Construction sites are potentially dangerous in many ways, but only those dangers that are associated with the risks of electric shock or burns are considered here. The risk of electric shock is high on a construction site because:

1 of the possibility of damage to cables and equipment
2 of the wide use of hand tools with trailing leads
3 of the accessibility of many extraneous-conductive-parts, which cannot practically be bonded
4 the works are generally open to the elements.

704.410.3.5
704.410.3.6

Section 704 prohibits the protective measures of obstacles and placing out of reach (Section 417), non-conducting location (Regulation 418.1), and earth-free local equipotential bonding (Regulation 418.2).

4.4 Supplies

704.411.3.1

Regulation 704.411.3.1 states that a PME earthing facility shall not be used for the means of earthing for an installation falling within the scope of this section unless all extraneous-conductive-parts are reliably connected to the main earthing terminal in accordance with Regulation 411.3.1.2. A note in the regulation also refers the reader to BS 7375 (Code of practice for distribution of electricity on construction and building sites).

It is usually impracticable with PME to comply with the bonding requirements of the Electricity Safety, Quality and Continuity Regulations on construction sites, and a PME earthing terminal should not be provided.

Where a construction site is part of an existing building and the building is supplied using a PME earthing facility the construction site will have to be separated from the PME building earth and be part of a TT system having a separate connection to earth independent from the PME building earth, unless the bonding requirements of Regulation 704.411.3.1 can be met. Where the construction site earthing and the building earthing are separate the earthing arrangements must not be simultaneously accessible.

The work should be planned in accordance with the Construction (Design and Management) Regulations 2007.

4.5 Protection against electric shock

4.5.1 Supplies to socket-outlets and hand-held equipment

Supplies to socket-outlets, etc must comply with the requirements of Regulation 704.410.3.10:

704.410.3.10

704.410.3.10 A circuit supplying a socket-outlet with a rated current up to and including 32 A and any other circuit supplying hand-held electrical equipment with rated current up to and including 32 A shall be protected by:

(i) reduced low voltage (Regulation 411.8), or

(ii) automatic disconnection of supply (Section 411) with additional protection provided by an RCD having the characteristics specified in Regulation 415.1.1, or

(iii) electrical separation of circuits (Section 413), each socket-outlet and item of hand-held electrical equipment being supplied by an individual transformer or by a separate winding of a transformer, or

(iv) SELV or PELV (Section 414).

Where electrical separation is used, special attention should be paid to the requirements of Regulation 413.3.4.

BS 7671 strongly prefers the reduced low voltage system to supply portable hand lamps for general use and portable hand tools and local lighting up to 2 kW, while SELV is strongly preferred for portable hand lamps in confined or damp locations.

411.8   **4.6   Reduced low voltage**

110 V reduced low voltage supplies, with the centre point of the secondary winding of the step-down transformer earthed, limit the voltage to Earth to 55 volts for single-phase supplies and 63.5 volts for three-phase.

Limiting the voltage to 55 volts or 63.5 volts between a live conductor and Earth effectively eliminates the risk of dangerous electric shock to exposed-conductive-parts. Figure 4.1 shows single-phase and three-phase reduced low voltage supplies.

▼ **Figure 4.1**
Reduced low voltage
supplies

a) Single-phase

b) Three-phase

704.511.1   The movable reduced low voltage 110 volt installation equipment is required to comply with BS EN 60439-4. 110 V plugs and sockets to BS EN 60309-2 (or BS 4343) are coloured yellow as recommended by the code of practice for distribution of electricity on construction and building sites (BS 7375). It is usual practice for the cables in the movable installation to have yellow sheaths to identify the voltage being used, although this is not a requirement of BS 7671 or BS 7375. Cable manufacturers produce arctic grade yellow PVC sheathed flexible cables for use in low temperature situations to distinguish from normal PVC sheathed cables.

It should be noted that whilst BS 7671 strongly prefers reduced low voltage systems for portable hand lamps for general use and portable hand tools and local lighting up to 2 kW, the Health and Safety Executive have advised that the use of 230 V hand tools individually protected by an individual 30 mA RCD may be acceptable

The notes of guidance to the Electricity Safety, Quality and Continuity Regulations advise that special consideration be given to the earthing and protection arrangements. GN5 Guidance Note 5: *Protection Against Electric Shock* and Guidance Note 8: *Earthing &* GN8 *Bonding* include guidance on earthing.

In addition to reduced low voltage, BS 7375 also provides information on typical distribution voltages for particular applications. For example, 400 V three-phase for fixed or movable plant over 3.75 kW and 230 V single-phase for fixed floodlighting and site building installations.

4.6.1 Maximum earth fault loop impedances

411.8.3 The value of earth fault loop impedance at every point of utilisation, including socket-outlets, must be such that the disconnection time does not exceed 5 s.

The maximum value of earth fault loop impedance is determined by the formula:

$$Z_S \times I_a \leq U_0$$

where:

Z_S is the earth fault loop impedance
I_a is the current in amperes causing automatic operation of the device within the specified time
U_0 is the nominal line voltage to Earth in volts.

Alternatively, if the nominal voltage U_0 is 55 V single-phase or 63.5 V three-phase then Table 41.6 of BS 7671:2008 can be used to determine the maximum earth fault loop impedances for various circuit-breaker types and ratings. However, it should be noted that the values given are not to be exceeded when the conductors are at their normal operating temperature.

If general-purpose fuses to BS 88-2 are used and the nominal voltage U_0 is 55 V single-phase or 63.5 V three-phase then Table 41.6 of BS 7671:2008 can again be used to determine the applicable maximum earth fault loop impedances.

For a fuse of a different type and rating, the appropriate British Standard should be referenced to determine the value of I_a for the disconnection time in accordance with the appropriate value of the nominal voltage U_0. This can then be used to determine the maximum earth loop impedance to comply with $Z_S \times I_a \leq U_0$.

4.7 Wiring systems

704.52 Cables on a construction site location should preferably not be installed across walkways or site roads as they are susceptible to mechanical damage. If cables are installed in this manner they would require the appropriate level of mechanical protection.

For reduced low voltage systems flexible thermoplastic cables rated at 300/500 V and suitable for low temperature (BS 7919) should be used. These cables remain flexible at lower temperatures than standard PVC cables and are ideal for outdoor use. They are referred to as arctic grade cable and typically have yellow (refer to section 4.5) or blue sheaths.

For cables used for applications exceeding reduced low voltage, flexible cables rated at 450/750 V that are resistant to abrasion and water should be used, type H07RN-F (BS 7019) or equivalent. These are heavy duty rubber insulated and sheathed flexible cables suitable for outdoor use.

4.8 Isolation and switching

704.537.2.2 Regulation 704.537.2.2 requires each Assembly for Construction Sites (ACS) to incorporate suitable devices for the switching and isolation of the incoming supply.

A device for isolating the incoming supply must be suitable for securing in the off position, e.g. by padlock or location of the device inside a lockable enclosure.

Current-using equipment must be supplied by ACSs (see Figure 4.2), each ACS comprising:

i   overcurrent protective devices, and
ii devices affording fault protection, and
iii socket-outlets, if required.

Safety and standby supplies must be connected by means of devices arranged to prevent interconnection of the different supplies.

▼ **Figure 4.2**
Examples of assemblies for construction sites

a Typical power assembly   **b** Typical lighting assembly

4.9 Protection against the weather and dust

All equipment that is part of the movable installation should have a degree of protection appropriate to the external influences. Equipment for external use should be at least IP44. However, equipment installed in a weather protected location, such as an office being refurbished, would have no specific IP requirement.

4.10 Inspection and testing

GN3 It is recommended that the maximum period between inspections of construction site installations is 3 months.

Fixed installation RCDs should additionally be tested daily (using the integral test button). Should RCDs be used to protect mobile equipment they must be tested by the operative before each period of use (again using the integral test button) and by the responsible person every 3 months (using an RCD tester).

Recommended intervals for inspection and testing are given in Table 4.1.

▼ **Table 4.1** Frequency of inspection and testing of equipment on construction sites

Type of equipment (note 1)	User checks (note 2)	Class I		Class II (note 4)	
		Formal visual inspection (note 3)	Combined inspection and testing (note 5)	Formal visual inspection (note 3)	Combined inspection and testing (note 5)
S	None	1 month	3 months	1 month	3 months
IT	None	1 month	3 months	1 month	3 months
M (note 6)	weekly	1 month	3 months	1 month	3 months
P (note 6)	weekly	1 month	3 months	1 month	3 months
H (note 6)	weekly	1 month	3 months	1 month	3 months

Notes:

1 S Stationary equipment
   IT Information technology equipment
   M Movable equipment
   P Portable equipment
   H Hand-held equipment
2 User checks are not recorded unless a fault is found.
3 The formal visual inspection may form part of the combined inspection and tests when they coincide, and must be recorded.
4 If class of equipment is not known, it must be tested as Class I.
5 The results of combined inspections and test are recorded.
6 30 mA RCDs supplying this equipment should be tested daily using the integral test button. Before carrying out the test it should be confirmed that it is safe to do so.

Agricultural and horticultural premises

5

5.1 Introduction

Introduction to Amendment No 1 to the 17th Edition

Sect 705   Section 705 of Amendment No 1 to BS 7671:2008 includes some notable changes.

705.411.1   Under the protective measure 'Automatic disconnection of supply', indent (i) of Regulation 705.411.1 has been modified making reference to an RCD having the characteristics specified in Regulation 415.1.1. Editorial changes have been made to Regulations 705.411.4, 705.414.4.5 and 705.415.2.1. Also, under Selection and erection of equipment, previous Regulation 705.559 Luminaires and lighting installations has been deleted.

The main changes introduced by the 17th Edition (BS 7671:2008) when it was published in January 2008 are described below for information.

Introduction to the 17th edition

Sect 705   Section 705 'Agricultural and horticultural premises' in the 17th Edition has been restructured and there are a number of editorial and technical changes.

For automatic disconnection of supply the 0.2 s maximum disconnection time has been removed and replaced by the requirement for the use of RCDs as disconnection devices for circuits, irrespective of the type of earthing system.

705.415.2.1   Regulation 705.415.2.1 expands the requirements of Regulation 415.2 concerning supplementary equipotential bonding and refers to Figure 705 of BS 7671:2008 at the end of the section showing the requirements for bonding in cattle sheds. A note to this regulation indicates that TN-C-S is not recommended if there is no metal grid laid in the floor. (TN-C-S now changed to PME by Amendment No 1.)

Under 'Safety services', a new requirement in Regulation 705.560.6 requires consideration of automatic life support in high density livestock environments, to ensure that supplies of water, food, ventilation and lighting are maintained in the event of power failure by means of, for example, an alternative source or back-up supply.

5.2 Scope

The particular requirements of Section 705 apply to fixed electrical installations indoors and outdoors in agricultural and horticultural premises. Some of the requirements are also applicable to other locations that are in common buildings belonging to the agricultural and horticultural premises. Where special requirements also apply to residences and other locations in such common buildings this is stated in the text of the relevant regulations.

Rooms, locations and areas for household applications and similar are not covered by this section.

Note: Chapter 5 does not cover electric fence installations, for which reference should be made to BS EN 60335-2-76.

5.2.1 Residences and other locations belonging to agricultural and horticultural premises

Part 2   These are defined as:

Residences and other locations which have a conductive connection to the agricultural and horticultural premises by either protective conductors of the same installation or by extraneous-conductive-parts.

Note: Examples of other locations include offices, social rooms, machine-halls, workrooms, garages and shops.

5.3   The risks

The particular risks associated with farms and horticultural premises are:

1   general accessibility of extraneous-conductive-parts
2   an onerous environment with respect to mechanical damage, exposure to the weather, corrosive effects (from water, animal urine, farm chemicals etc.)
3   a mechanically hazardous area due to electromechanical equipment, mills and mixers, and mechanical drives of all kinds
4   rodent damage to (gnawing of) cables, leading to fire risks
5   storage of flammable materials, e.g. straw and grain
6   increased susceptibility of electric shock for livestock.

5.4   Electricity supplies

Because of the practical difficulties in bonding all accessible extraneous-conductive-parts electricity distribution companies might not provide a PME earth to agricultural or horticultural installations.

The Department of Energy and Climate Change (DECC), formerly the DTI, guidance on the Electricity Safety, Quality and Continuity Regulations 2002 advises {9(4)} that special consideration should be given to the earthing and bonding requirements to farms where it may prove difficult to attach and maintain all the necessary equipotential bonding connections for a PME supply.

705.415.2.1   Regulation 705.415.2.1 refers to Figure 705, which shows the requirements for bonding in cattle sheds. PME is **not** recommended if there is no metal grid laid in the floor.

A TN-S supply is unlikely to be provided by a distributor as a routine unless the installation is particularly large and early application is made. Alternatively, consideration should be given to installing an additional earth electrode, as it is most likely that the installation will be required to be TT.

It should be noted that in TT installations isolators are required to switch all live conductors including the neutral.

705.411.4   Regulation 705.411.4 prohibits the use of a TN-C system.

5.5 Protection against electric shock

705.410.3.5 The protective measures of obstacles and placing out of reach (Section 417) are not permitted.

705.410.3.6 The protective measures of non-conducting location (Regulation 418.1) and earth-free local equipotential bonding (Regulation 418.2) are not permitted.

5.5.1 Protective measure: automatic disconnection of supply

705.411.1 RCDs must be provided for automatic disconnection of supply to circuits as follows (see Figure 5.1):

> **705.411.1** In circuits, whatever the type of earthing system, the following disconnection devices shall be provided:
>
> 415.1.1
> (i) In final circuits supplying socket-outlets with rated current not exceeding 32 A, an RCD having the characteristics specified in Regulation 415.1.1
> (ii) In final circuits supplying socket-outlets with rated current more than 32 A, an RCD with a rated residual operating current not exceeding 100 mA
> (iii) In all other circuits, RCDs with a rated residual operating current not exceeding 300 mA.

▼ **Figure 5.1**
Requirements for RCDs

Notes:
1 Socket-outlet circuits not exceeding 32 A
2 Socket-outlet circuits exceeding 32 A
3 Circuits other than socket-outlets
4 The risk of faults to exposed-conductive-parts on the supply side of the main switch must be minimised as such faults are not detected by the RCDs.

PD 6519-3:1999 (IEC 60479-3:1998) A BSI published document PD 6519-3 (IEC 60479-3) *Guide to effects of current on human beings and livestock – Part 3 Effects of current passing through the body of livestock* is available that provides guidance on the effects of electric currents on livestock. The report indicates values for the impedance of the body of livestock as a function of the touch voltage, the degree of moisture of the hide or skin and the current path.

The report includes the following information:

▶ The internal body impedance is considered as mostly resistive and depends primarily on the current path.
▶ The impedance of a hide depends largely on humidity. In dry conditions the hide can be considered to be practically an insulator for voltages up to 100 V with impedance values in the range of tens to hundreds of kilohms.
▶ The impedance of skin depends on the voltage, frequency, duration of current flow, surface area of contact, pressure of contact, degree of moisture of the skin and temperature. The skin impedance falls as the current is increased.
▶ The impedance of the hind legs is smaller than the impedance of the forelegs. The impedance from the nose to all four legs is smaller than the impedance of forelegs to hind legs.
▶ The total body impedance (cattle) for a current path nose to the four legs is typically 35% of the total body impedance for a current path forelegs to hindlegs.

However, animals will detect relatively small voltage gradients between front and rear legs, and between conductive part potentials and Earth.

705.415.2.1 This can result in, for example, a marked reluctance for cows to enter milking parlours because of potential differences. These potential differences can arise from a number of causes. If the electricity supply is PME, at the end of a long run there is likely to be a potential between the PME earth and true Earth. In locations intended for livestock, supplementary bonding must connect all exposed-conductive-parts and extraneous-conductive-parts that can be touched by livestock. Where a metal grid is laid in the floor, it must be included within the supplementary bonding of the location (Figure 705 of BS 7671 shows an example of this; other suitable arrangements of a metal grid are not precluded). Extraneous-conductive-parts in, or on, the floor, e.g. concrete reinforcement in general or reinforcement of cellars for liquid manure, must be connected to the supplementary equipotential bonding. It is recommended that spaced floors made of prefabricated concrete elements be part of the supplementary equipotential bonding. The supplementary equipotential bonding and the metal grid, if any, must be erected so that it is durably protected against mechanical stresses and corrosion.

Note: Where a metal grid is not laid in the floor a PME supply is not recommended.

It is well known that animals have received shocks that are associated with electrical installations. It is likely that lightning strikes on overhead lines conducted to Earth via earth electrodes at the bottom of a pole produce voltage gradients that are fatal to animals because of the wide spacing of their feet. If there is concern in this respect, the location of earth electrodes should be discussed with the electricity distribution company.

5.6 Earth fault loop impedances

411.5.3 Where an RCD is used for earth fault protection, the following conditions are to be fulfilled for a TT installation:

The disconnection time must satisfy Regulation 411.3.2.2 or 411.3.2.4, and
$R_A \times I_{\Delta n} \leq 50$ V

where:

R_A is the sum of the resistances of the earth electrode and the protective conductor connecting it to the exposed-conductive-parts (in ohms)

$I_{\Delta n}$ is the rated residual operating current of the RCD.

The above conditions would be met where the earth fault loop impedance does not exceed the values stated in Table 5.1.

▼ **Table 5.1**
Maximum earth fault
loop impedance

Rated residual operating current, $I_{\Delta n}$ (mA)	Maximum earth fault loop impedance, Z_S (Ω)
30	1667*
100	500*
300	167

* The resistance of the installation earth electrode itself should be as low as practicable. A value exceeding 200 ohms may not be stable. Refer to Regulation 542.2.4.

5.7 Protection against fire

Regulation 705.422.6 requires electrical heating appliances used for the breeding and rearing of livestock to comply with BS EN 60335-2-71 and to be fixed so as to maintain an appropriate distance from livestock and combustible material, to minimise any risks of burns to livestock and of fire. For radiant heaters the clearance should be not less than 0.5 m or such other clearance as recommended by the manufacturer. For luminaires refer to Regulations 422.3.1 and 422.4.2.

705.422.7 Regulation 705.422.7 requires that for fire protection purposes, RCDs are to be installed with a rated residual operating current not exceeding 300 mA and must disconnect all live conductors. Where improved continuity of service is required, RCDs not protecting socket-outlets need to be of the S type or have a time delay (see Figure 5.2).

A note to Regulation 705.422.7 advises that the protection of final circuits by RCD required according to Regulation 411.1 is also effective for protection against fire.

705.422.8 Regulation 705.422.8 specifies requirements for conductors of circuits supplied at extra-low voltage in locations where a fire risk exists.

Rodent damage is a major cause of farm fires and this must be taken into account by the designer and installer. Cables should be installed and routed with such potential damage in mind. For example, in a livestock building, cables should be routed on the underside of the ceiling rather than in a false roof. Steel conduit provides a good degree of protection.

GN4 Additional guidance for protection against fire for this type of location can be found in section 8.2 of Guidance Note 4: *Protection Against Fire*.

▶ **Figure 5.2** Example schematic of supply arrangements (including livestock support)

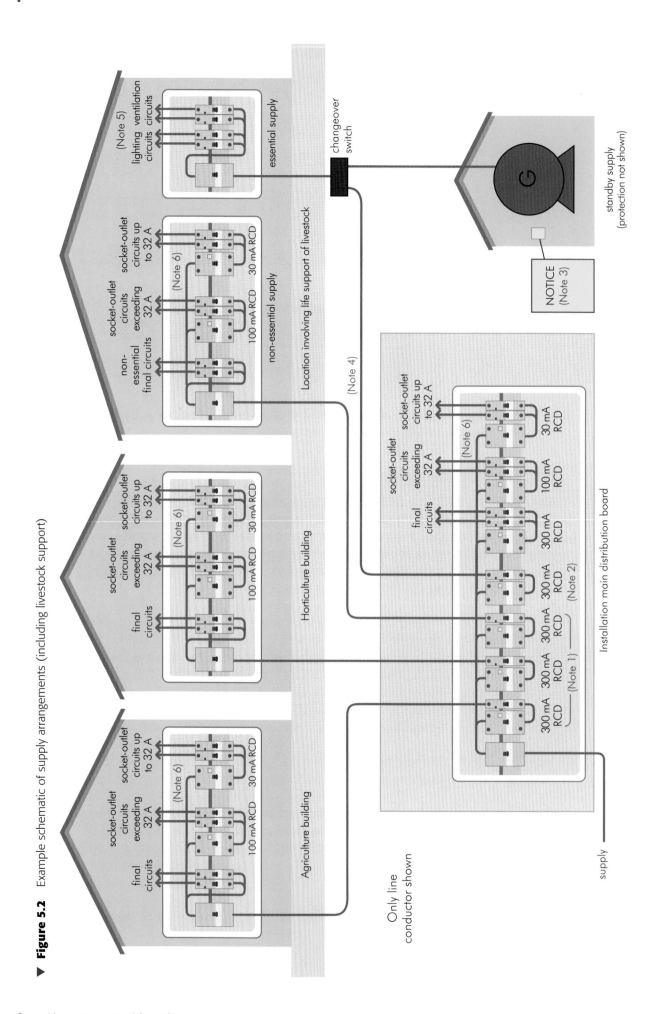

Notes to Figure 5.2:

1 All circuits other than socket-outlets must be protected by 300 mA RCDs for fire protection purposes. These will need to discriminate with final circuit RCDs where relevant.

2 For improved continuity of service, RCDs should be time delay or S type as long as they are not supplying socket-outlets.

3 A standby supply should have a notice adjacent to highlight that it should be periodically tested in accordance with manufacturer's instructions.

4 Discrimination of main circuits supplying ventilation for life support circuits.

5 Separation of lighting and ventilation circuits for life support.

6 Circuits supplying socket-outlets up to 32 A require protection by 30 mA RCDs. Circuits supplying socket-outlets exceeding 32 A require protection by 100 mA RCDs.

5.8 External influences

705.512.2 Electrical equipment is required to have a minimum degree of protection of IP44, when used under normal conditions. Where equipment of IP44 rating is not available, it should be placed in an enclosure complying with IP44.

▶ IPX4 provides protection against water splashing.
▶ IPX5 provides protection against water jets from any direction.
▶ IPX6 provides protection against powerful water jets from any direction.

GN1 See Appendix B of Guidance Note 1: *Selection & Erection* for details of the IP code and the equivalent (now superseded) drip-proof, etc. symbols.

Socket-outlets should be installed in a position where they are unlikely to come into contact with combustible material. Where there are conditions of external influences >AD4, >AE3 and/or >AG1, socket-outlets must be provided with the appropriate protection. Protection may also be provided by the use of additional enclosures or by installation in building recesses.

See Appendix 5 of BS 7671:2008 for classification of external influences.

The foregoing requirements do not apply to residential locations, offices, shops and locations with similar external influences belonging to agricultural and horticultural premises where, for socket-outlets, BS 1363-2 or BS 546 applies. Where corrosive substances are present, e.g. in dairies or cattle sheds, the electrical equipment needs to be adequately protected.

705.559 Luminaires should comply with the BS EN 60598 series and be selected regarding their degree of protection against the ingress of dust, solid objects and moisture (e.g. IP54).

Luminaires marked ▽D̲ have a limited surface temperature of the luminaire, and should have a degree of protection of IP54.

5.9 Wiring systems

_{705.522} Regulation 705.522 calls for wiring systems to be inaccessible to livestock or suitably protected against mechanical damage.

Where vehicles and mobile agricultural machines are operated, underground cables must be buried in the ground at a depth of at least 0.6 m with added mechanical protection. Cables in arable or cultivated ground should be buried at a depth of at least 1 m. Self-supporting suspension cables should be installed at a height of at least 6 m.

_{705.522.16} Regulation 705.522.16 has requirements for conduit, cable trunking and ducting systems to be protected against corrosion and impact.

5.10 Safety services

_{705.560.6} Where an installation includes high density livestock rearing, there may be a need to take account of continued operation of systems for the livestock (see Figure 5.2).

If the supply of food, water, air and/or lighting to the livestock cannot be ensured in the event of a power supply failure then a separate source of supply should be provided. This would include either an alternative supply or back-up supply.

For the supply of ventilation and lighting units, separate final circuits should be provided. These circuits should be limited to support just the essential equipment of the ventilation and lighting. Discrimination of the circuits supplying ventilation should be provided.

Electrically powered ventilation should include either of the following:

▶ A standby electrical source of supply that can ensure the operation of the ventilation equipment. If a standby source is used then a notice adjacent to this source should be provided highlighting that the standby source should be tested periodically according to the manufacturer's instructions.
▶ Monitoring of the temperature and supply voltage by one or more devices. These should provide a visual or audible signal that is located in a position that can be readily observed by the user, and the device(s) should operate independently from the normal supply.

Conducting locations with restricted movement

6

6.1 Introduction

Introduction to Amendment No 1 to the 17th Edition

The scope now includes the following statement:

Sect 706 This section does not apply to electrical systems as defined in BS 7909 used in structures, sets, mobile units etc as used for public or private events, touring shows, theatrical, radio, TV or film productions and similar activities of the entertainment industry.

The changes introduced by the 17th Edition (BS 7671:2008) when it was published in January 2008 have been mentioned below for information.

Introduction to the 17th Edition

The section on conducting locations with restricted movement (titled 'Restrictive conductive locations' in the 16th Edition) is retained. The only notable change in the 17th Edition is the addition of PELV to the list of suitable protective measures, subject to the necessary bonding being in place.

6.2 Scope

Part 2
706.1 A *conducting location with restricted movement* is constructed mainly of metallic or other conductive surrounding parts, and within it movement is restricted. It is likely that a person in such a location will be in good contact with conductive surroundings and escape will be difficult in the event of an electric shock. The regulations apply to the fixed conducting location and to the supplies for mobile and hand-held equipment for use in such locations.

The particular requirements of Section 706 do not apply to locations which allow persons freedom of bodily movement, that is, to enter, work and leave the location without physical constraint. The types of location that are being considered include boiler shells, cable gantries, small tunnels, metal sewers, etc.

6.3 The risks

In many ordinary locations, there is usually limited access to earthed metal. As a result, the likelihood of receiving a shock current of sufficient magnitude to have harmful physiological effects is low. This is not so with a conducting location.

In a conducting location where bodily movement is limited, there is little opportunity to move away from the shock. Contact resistance is low due to high contact areas and perspiration, so that body currents are high and the risk of ventricular fibrillation is also high. There are other effects of electric shock that are also relevant in such locations. Muscles used to breathe can be constrained by body currents, and shocks to the head can paralyse the breathing function.

6.4 Protection against electric shock

706.410.3.5 The protective measures of obstacles and placing out of reach are not permitted in these locations.

706.410.3.10 For supplies to equipment, the following protective measures are permissible:

Hand-held tools or items of mobile equipment
a electrical separation, only one item of equipment being connected to a secondary winding of the transformer (which may have two or more such windings), or

b SELV.

Handlamps
a SELV

Fixed equipment
a automatic disconnection of supply with supplementary equipotential bonding connecting exposed-conductive-parts of fixed equipment and the conductive parts of the location, or

b use of Class II equipment or equipment having equivalent insulation, provided the supply circuits have additional protection by the use of an RCD with a rated residual operating current not exceeding 30 mA having the characteristics specified in Regulation 415.1.1, or

415.1.1

c electrical separation, only one item of equipment being connected to a secondary winding of the transformer, or

d SELV or PELV.

For PELV, equipotential bonding must be provided between all exposed-conductive-parts, extraneous-conductive-parts inside the location and the connection of the PELV system to Earth.

706.413 For electrical separation, in addition to the requirements of Section 413, the unearthed source should have simple separation and should be situated outside the conducting location with restricted movement, unless the source is part of the fixed electrical installation of the location.

706.414 For SELV or PELV, the general requirements of Section 414 apply, except that, whatever the nominal voltage, basic protection must be provided. In addition, the source of SELV or PELV must be situated outside the conducting location with restricted movement, unless it is part of the fixed installation.

Electrical installations in caravan/camping parks, caravans and motor caravans

7

7.1 Introduction

Introduction to Amendment No 1 to the 17th Edition

Sect 708   **Section 708**

708.411.4   Under 'Protection against electric shock', TN system, Regulation 708.411.4 has been modified, and now refers to PME.

708.553.1.12   Regulation 708.553.1.12 now includes overcurrent requirements for a supply to each leisure accommodation vehicle or tent, whether via a socket-outlet or fixed connection.

708.553.1.13   Regulation 708.553.1.13 now includes RCD requirements for a supply to each leisure accommodation vehicle or tent, whether via a socket-outlet or fixed connection.

708.553.1.14   Regulation 708.553.1.14 now refers to PME. Finally, the note on identification (Fig 708) has been modified.

Sect 721   **Section 721**

The requirements concerning proximity to non-electrical services regarding cables and electrical equipment in a gas cylinder storage compartment has been revised. Also identification requirements have changed. In addition the guidance in the annex in connection with proximity of wiring systems with other services has changed.

Introduction to the 17th Edition

The 16th Edition included two separate divisions within the same section. The first of these addressed caravans and motor caravans, the second addressed caravan parks.

Sect 708
Sect 721   The 17th Edition has two separately numbered sections, namely Section 708 concerning the electrical installations in caravan parks, etc., and Section 721 concerning caravans and motor caravans.

This chapter provides guidance on applying the particular requirements of both Sections 708 and 721 of BS 7671:2008.

7.2 Scope of Sections 708 and 721

7.2.1 Section 708

The particular requirements of this section apply to that portion of the electrical installation in caravan/camping parks and similar locations providing facilities for supplying leisure accommodation vehicles (including caravans) or tents.

The requirements do not apply to the internal electrical installations of leisure accommodation vehicles or mobile or transportable units.

7.2.2 Section 721

The particular requirements of this section apply to the electrical installations of caravans and motor caravans at nominal voltages not exceeding 230/440 V a.c. or 48 V d.c., except for 12 V d.c. – see exclusions below.

The requirements do not apply to the following:

▶ electrical circuits and equipment intended for the use of the caravan for habitation purposes
▶ electrical circuits and equipment for automotive purposes
▶ 12 V d.c. installations, which are covered by BS EN 1648-1 and BS EN 1648-2
▶ electrical installations of mobile homes or residential park homes, to which the general requirements apply
▶ transportable units (see Chapter 14 of this Guidance Note).

For the purposes of Section 721, caravans and motor caravans are referred to as 'caravans'. The particular requirements of some other sections of Part 7 may also apply to electrical installations in caravans, e.g. Section 701 (showers).

7.3 The risks

The risks specifically associated with installations in caravan parks, caravans and motor caravans arise from:

i open circuit faults of the PEN conductor of PME supplies raising the potential to true Earth of all metalwork (including that of caravans if connected) to dangerous levels
ii incorrect polarity at the pitch supply point
iii inability to establish an equipotential zone external to the vehicle
iv possible loss of earthing due to long supply cable runs, connecting devices exposed to weather and flexible cable connections liable to mechanical damage
721.522.7.1 **v** vibration while the vehicle is moving, causing faults within the caravan installation.

Particular requirements to reduce the above risks include:

708.411.4 **1** prohibition of the connection of exposed- and extraneous-conductive-parts of a
708.553.1.14 caravan or motor caravan to a PME terminal. The protective conductor of each socket-outlet to be connected to an earth electrode and the requirements of
411.5 Regulation 411.5 for a TT system to be complied with
721.411.1 **2** additional protection by 30 mA RCDs in both the vehicle and the park
708.553.1.13 installation
721.537.2.1.1 **3** double-pole isolating switch and final circuit circuit-breakers in the caravan or
721.43.1 motor caravan
721.524.1 **4** internal wiring of the caravan or motor caravan by flexible or stranded cables of
721.522.8.1.3 cross-sectional area 1.5 mm² or greater; additional cable supports; segregation of
721.528.1 low voltage and extra-low voltage circuits.

7.4 Legislation and standards

Regulation 9(4) of the Electricity Safety, Quality and Continuity Regulations 2002 prohibits the connection of the supply neutral of a PME supply to any metalwork in a caravan (or boat).

Caravan parks (sites) in the United Kingdom are subject to the provisions of the Caravan Sites and Control of Development Act 1960. This empowers local authorities to issue licences and to impose conditions, generally in accordance with model standards. For residential parks there are the *Model Standards 2008 for Caravan Sites in England* and *Model Standards 2008 for Caravan Sites in Wales*. For holiday sites there are the *Model Standards 1989: Holiday Caravan Sites*. For touring sites there are the *Model Standards for Touring Caravan Sites*. These model standards include requirements for the electrical installation in the caravan park to be installed and maintained to the requirements of the current edition of the *IET Wiring Regulations* (BS 7671).

The 1960 Act also empowers certain 'exempted bodies' such as the Caravan Club and the Camping and Caravanning Club to issue certificates in respect of parks for use by their own members.

While there is no legislation specific to electrical installations in caravans, BS EN 1645-1 *Leisure accommodation vehicles – caravans* includes a section on electrical installations. This requires that low voltage electrical installations are to comply with international standard IEC 60364-7-708 *Low voltage electrical installations – Part 7-708: Requirements for special installations or locations – Caravan parks, camping parks and similar locations*. These requirements are included in Section 708 of BS 7671:2008.

Extra-low voltage installations using 12 V d.c. should comply with BS EN 1648-1 *Leisure accommodation vehicles – 12 V direct current extra low voltage electrical installations – Part 1: Caravans*.

7.5 Electrical installations in caravan parks (708)

7.5.1 Supply systems

708.411.4 The supply system to the permanent buildings of a caravan park can be PME, TN-S or TT. Consideration should be given to the earthing and bonding of amenity buildings such as toilet and shower blocks. The supply system to the caravans is limited to TN-S or TT. The Electricity Safety, Quality and Continuity Regulations (ESQCR) prohibit the connection of a PME earthing facility to any metalwork in a leisure accomodation vehicle (caravan). If the caravan supply is derived from a permanent building that is supplied by a PME system then the caravan supply will have to be part of a TT system having a separate connection to Earth independent from the PME earthing.

542.1.8 The separation of the earthing should preferably be effected at the main distribution board (see Figure 7.1). This enables the exposed-conductive-parts connected to each system to be more readily identified and inspected periodically. An earth electrode for the TT system should be provided nearby and located so that the resistance areas of the PME supply earthing and earth electrode do not overlap (refer to section 7.5.5).

Alternatively, the separation of the earthing can be made at the caravan pitch supply points. In this instance, earth electrodes will be required at these points. A cable supplying a separate earthing system is to be earthed only at the installation containing the associated protective device (see Figure 7.2).

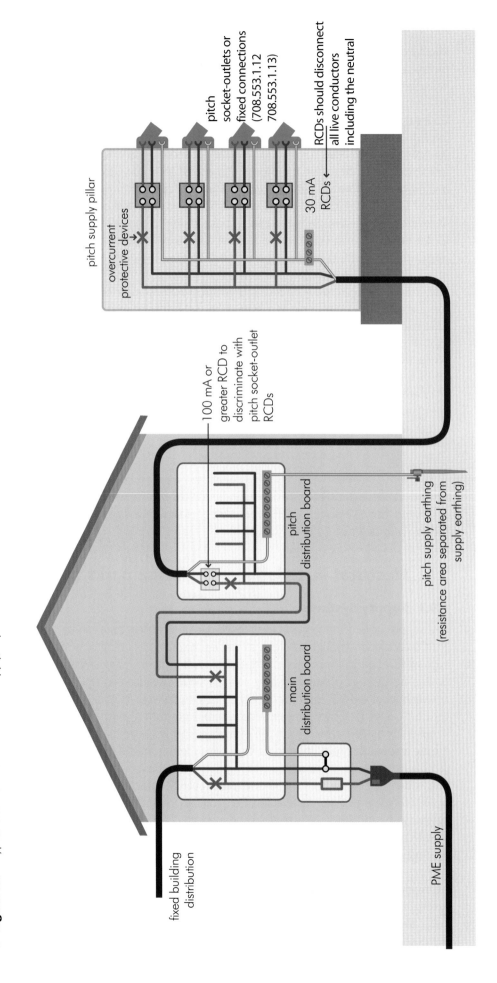

▶ **Figure 7.1** Typical site distribution for a PME supply, separation from PME earth at main distribution board

▼ **Figure 7.2** Typical site distribution for a PME supply, separation from PME earth at pitch supply point

pitch
socket-outlets or
fixed connections
(708.553.1.12
708.553.1.13)

RCDs should disconnect
all live conductors
including the neutral

pitch supply pillar

overcurrent
protective devices

30 mA
RCDs

pitch
supply earthing

c.p.c. of
distribution cable
segregated from
pitch supply
earthing

distribution board

fixed building
distribution

PME supply

411.3.2.2
411.3.2.3
411.3.2.4
314.1(i)

For automatic disconnection of supply, final circuits not exceeding 32 A require maximum disconnection times of 0.4 s for TN and 0.2 s for TT systems. Final circuits exceeding 32 A and distribution circuits require maximum disconnection times of 5 s for TN and 1 s for TT systems. For TT supplies an RCD of 100 mA or more will be required which will need to be a delay or 'S' type to discriminate with the individual 30 mA RCDs required for the pitch outlets. If an RCD is used for a TN-S system then discrimination will again be required.

7.5.2 Overhead and underground distribution

708.521.1

708.410.3.5

BS 7671 states a preference for distribution by underground cable. Underground cables should be buried at a depth of at least 0.6 m and should be routed outside any caravan pitch or away from areas where tent poles or ground anchors could be used, unless additional mechanical protection is provided. Overhead distribution systems are allowed provided that all conductors are insulated and so erected as to be unlikely to be damaged by vehicle movement. They are required to be not less than 6 m above ground level in areas subject to vehicle movement and 3.5 m in other areas. Suitable warning notices should be displayed at the entrance to the site and on supports for the overhead line and, where appropriate, attention drawn to the danger of masts of yachts or dinghies contacting the overhead line.

7.5.3 Caravan pitches

Dimensions of pitches are provided within the model standards issued under the Caravan Sites and Control of Development Act 1960, Section 5. Dimensions of pitch relate to the type of site as follows:

▶ **For a holiday site.** There should be a 3 m wide area kept clear within the inside of all boundaries. For units made of aluminium or similar fire performance properties, there should be a minimum of 5 m between units (3.5 m at the corners). For units made of plywood or similiar, there should be a minimum of 6 m. For a mixture of types a minimum of 6 m. For a mixture of residential homes and holiday caravans a minimum of 6 m.

▶ **For a touring site.** There should be a 6 m minimum distance between units in separate family occupation and 3 m minimum between units in any circumstances.

▶ **For a residential site.** There should be a 6 m minimum distance between caravans. If a caravan has been fitted with cladding from Class 1 fire rated materials then the minimum distance is 5.25 m.

When designing pitches, the requirements of the above legislation should be confirmed with the relevant local authority.

7.5.4 Caravan pitch socket-outlets

708.553.1.8
708.553.1.10

708.530.3

Caravan pitch socket-outlets are required to comply with BS EN 60309-2 and to have a degree of protection of at least IP44. The current rating is to be not less than 16 A but may be greater if required. At least one socket-outlet should be provided for each caravan pitch. Where socket-outlets are grouped in pitch supply equipment, there should be one socket for each pitch limited to a group of four. To be compatible with the caravan connecting cable, sockets should be two-pole with earthing contact having key position 6 h.

708.553.1.12
708.553.1.13

Each socket-outlet must be protected individually by an overcurrent device, which may be a fuse but is more usually a circuit-breaker, and individually by an RCD having the characteristics specified in Regulation 415.1.1. The RCD must disconnect all live conductors including the neutral. See Figures 7.1 and 7.2.

7.5.5 Separation of electrodes

Figure 7.3 indicates that effective separation of resistance areas of earth electrodes is achieved if the distance between the electrodes exceeds 10 m.

▼ **Figure 7.3**
Ground surface potentials around a single rod and three rods in line

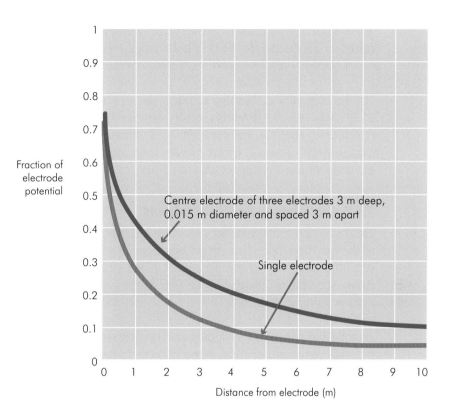

GN3 Guidance Note 3: *Inspection & Testing* describes a test method for measurement of earth electrode resistance.

7.6 Caravans and motor caravans (721)

7.6.1 Equipotential bonding

721.411.3.1.2 Structural metallic parts which are accessible from within the caravan are required to be connected through main protective bonding conductors to the main earthing terminal within the caravan.

7.6.2 Provision of RCDs

721.411.1 Where protection by automatic disconnection of supply is used (Section 411), an RCD complying with BS EN 61008-1 or BS EN 61009-1 interrupting all live conductors is required to be provided having the characteristics specified in Regulation 415.1.1 (30 mA), and the wiring system must include a circuit protective conductor connected to:

i   the protective contact of the inlet, and
ii the exposed-conductive-parts of the electrical equipment, and
iii the protective contacts of the socket-outlets.

7.6.3 Protection against overcurrent

721.43.1 Each final circuit must be protected by an overcurrent protective device which disconnects all the live conductors of the circuit.

7.6.4 Selection and erection of equipment

721.510.3 More than one electrically independent installation is permitted provided each one is supplied via a separate connecting device and they are segregated in accordance
528.1 with the requirements of Regulation 528.1. This includes requirements for insulating cables or conductors to the highest voltage present, or installing the independent circuit cables in separate conduits, trunking or ducting compartments.

This enables 12 V d.c. battery supplied circuits for interior lighting in accordance with BS EN 1648 *Leisure accommodation vehicles 12 V direct current extra low voltage installations, Part 1 – caravans, Part 2 – motor caravans* and for the road lighting circuits of the caravan or motor caravan in accordance with the correct amendment of the Road Vehicle Lighting Regulations 1989.

7.6.5 Switchgear and controlgear

721.537.2 The installation to the caravan should have a main disconnector which will disconnect all the live conductors. This should be placed in a suitable position ready for operation within the caravan to isolate the supply. When a caravan only has one final circuit then the isolation can be afforded by the overcurrent protective device as long as it fulfils the requirements for isolation.

An indelible notice in the appropriate language(s) must be permanently fixed near the main isolation point inside the caravan to provide the user with instructions on
Figure 721 connecting and disconnecting the supply (refer to Figure 721 of BS 7671).

721.55.1 The inlet to the caravan must be an appropriate inlet complying with BS EN 60309. This should be installed not more than 1.8 m above ground level, in a readily accessible position, have a minimum degree of protection of IP44, and should not protrude significantly beyond the body of the caravan.

7.6.6 The connecting flexible cable

721.55.2.6 The means of connection of the caravan to the pitch socket-outlet should be provided with the caravan. This must have a plug at one end complying with BS EN 60309-2, a flexible cable with a length of 25 m (±2 m) and a connector if needed that is compatible with the appropriate appliance inlet. The cable should be to the harmonized code H05RN-F (BS 7919) or equivalent, include a protective conductor, have cores coloured as required by Table 51 of BS 7671 and have a cross-sectional area as shown in Table 7.1.

▼ **Table 7.1**
Minimum cross-sectional areas of flexible cables for caravan connection (Table 721 of BS 7671:2008)

Rated current (A)	Minimum cross-sectional area (mm²)
16	2.5
25	4
32	6
63	16
100	35

7.7 Tents

708.553.1.12
708.553.1.13

Mains supplies to tents (marquees etc) should generally incorporate the equipment specified for a caravan, namely a main double-pole isolating switch and a main double-pole RCD (which may be combined), one or more double-pole circuit-breakers and socket-outlets as required. Double-pole circuit-breaker means having overcurrent detection and switching in both poles and not a single-pole circuit-breaker with a switched neutral. The connecting cable should be securely clamped to the main supply unit and directly connected to it without the use of a plug and inlet. Luminaires and appliances should be of Class II construction.

Marinas and similar locations 8

8.1 Introduction

Introduction to Amendment No 1 to the 17th Edition

709.411.4 Under 'Protection against electric shock', TN system, Regulation 709.411.4 has been modified and now refers to PME.

Cable management requirements under wiring systems have been changed slightly.

709.553.1.4 A new regulation, 709.553.1.14 has been introduced which prohibits the connection of a socket-outlet protective conductor to a PME earthing facility.

Introduction to the 17th Edition

Sect 709 The 16th Edition did not include requirements for marinas, but guidance was given in the previous edition of this Guidance Note. However, new Section 709 concerning electrical installations of marinas has been introduced into the 17th Edition.

Section 709 specifies particular requirements for this type of electrical installation relating to assessment of general characteristics, protection against electric shock and selection and erection of equipment.

For electrical installations of pleasure craft, reference should be made to BS EN 60092-507 *Electrical installations in ships – Pleasure craft*, while for houseboats the general requirements of BS 7671:2008 apply.

8.2 Scope

709.1 The particular requirements of Section 709 are applicable only to circuits intended to supply pleasure craft or houseboats in marinas and similar locations. (In this section 'marina' means 'marina and similar locations'.)

The particular requirements do not apply to the supply to houseboats if they are supplied directly from the public network, or to the internal electrical installations of pleasure craft or houseboats.

For the remainder of the electrical installation of marinas and similar locations the general requirements of the Regulations apply, together with any relevant particular requirements of Part 7.

8.3 The risks

709.512.2 The environment of a marina or yachting harbour is harsh for electrical equipment. The water, salt, and movement of structures accelerate deterioration of the installation. The presence of salt water, dissimilar metals and a potential for leakage currents increases the rate of corrosion. There are also increased electric shock risks associated with a wet environment, by reduction in body resistance and contact with Earth potential.

Site investigations should be carried out at an early stage to determine likely maximum wave heights. This is of particular importance in exposed coastal sites. Where marinas have breakwater type pontoons, it is likely that under certain conditions waves will pass over the structure.

The risks specifically associated with craft supplied from marinas include:

i open circuit faults of the PEN conductor of PME supplies raising the potential to true Earth of all metalwork (including that of the craft, if connected) to dangerous levels

ii inability to establish an equipotential zone external to the craft

iii possible loss of earthing due to long supply cable runs, connecting devices exposed to weather and flexible cable connections liable to mechanical damage.

Particular requirements to reduce the above risks include:

1 prohibition of the connection of exposed- and extraneous-conductive-parts of the craft to a PME terminal

2 additional protection by 30 mA RCDs in both the craft and the marina installation.

8.4 General requirements

Electrical power installations located at marinas should be installed and the equipment so selected as to minimise the risk of electric shock, fire and explosion. In the design and construction of such works, particular regard should be given to the risk of increased corrosion, movement of structures, mechanical damage, presence of flammable fuel and vapour and the physiological effects of electric shock being increased by a reduction in body resistance and contact of the body with earth potential.

Owing to the harsh working environment of marina installations and potential for abuse and accidental damage by users, particular attention should also be paid to the maintenance and periodic inspection reporting of installations and the general requirements of the Regulations.

8.5 Supplies

Where the supply system is protective multiple earthed (PME), Regulation 9(4) of the Electricity Safety, Quality and Continuity Regulations 2002 prohibits the connection of the neutral to the metalwork of any caravan or boat. While the PME supply may be fed to permanent buildings in the marina, supplies to boats (pleasure craft) must have a separate earth system. A TT system having a separate connection with Earth, independent of the PME earthing system (see Figure 8.3a/b and Figures 7.1 and 7.2), will meet this requirement. Alternatively, protection by electrical separation can be adopted; see section 8.7.

This avoids the risks arising from a loss of continuity of the supply PEN conductor.

The separation of the TT earthing system should be effected at the main distribution board, where the exposed-conductive-parts connected to each system can be more readily identified and inspected periodically. A main earth electrode for the TT system needs to be provided nearby, with no overlap of resistance area with any earthing
541.2 associated with the PME supply. (See also section 7.5.5 of Chapter 7.)

TN-S supplies may be made available both to permanent shore installations and to pleasure craft.

The nominal supply voltage of the installation to pleasure craft should not exceed 230 V single-phase or 400 V three-phase.

Some locations may have dry dock areas where construction and maintenance activities can be carried out. The guidance on supplies to these locations can be summarised as follows:

▶ For dry dock areas that form part of a formal workshop type environment under responsible control where construction and maintenance on vessels are carried out, reduced low voltage (110 V centre-tap earthed) and extra-low voltage fixed supplies should be provided. In addition, portable generators could be used that would require the appropriate earthing and protection.

▶ British Waterways have a number of dry dock locations accessible by vessel owners where maintenance activities can be carried out. Here, typical vessel distribution pillar fixed supplies with the appropriate overcurrent and RCD protection would be provided. The vessel owner would connect to this supply and use double insulated electrical equipment connected via the vessel socket-outlets.

▶ British Waterways also have a number of dry dock locations that do not have any fixed installation electrical supply. The supply in this type of location would be via a small generator. This would require the appropriate earthing and protection to be provided.

Note: Persons involved in designing and erecting electrical installations for dry dock facilities should contact the Health and Safety Executive and British Waterways for additional guidance.

8.6 Protection against electric shock

709.410.3 The protective measures of obstacles and placing out of reach are not permitted. Also, the protective measures of non-conducting location and earth-free local equipotential bonding are not permitted.

709.411.4 In the UK for a TN system, the final circuits for the supply to pleasure craft or houseboats shall not include a PEN conductor.

Note: In the UK the ESQCR prohibit the use of a PME system for the supply to a boat or similar construction.

Only permanent onshore buildings may use the electricity distributor's PME earthing terminal. For the boat mooring area of the marina this is not permissible, and entirely separate earthing arrangements must be provided. This is generally achieved by the use of a suitably rated RCD complying with BS EN 61008 with driven earth rods or mats providing a TT system for that part of the installation.

Marina installations are often of sufficient size to warrant the provision of an 11 kV/415 V transformer substation. In these, and sometimes in other, circumstances the electricity distributor may be willing to provide a TN-S supply, which is much more suitable for such installations. If the transformer belongs to the marina, a TN-S system should be installed.

8.7 Operational conditions and environmental factors

709.512.2.1.1
709.512.2.1.2
Electrical equipment to be installed on or above jetties, wharves, piers or pontoons must be selected according to the external influences which may be present. Regarding presence of solid foreign bodies, a minimum degree of protection of IP3X is required, and for presence of water the following applies.

External influence	Minimum index of protection
Presence of water splashes	IPX4
Presence of water jets	IPX5
Presence of waves of water	IPX6

709.512.2.1.3 In the marina environment, particularly at jetties, pontoons etc., consideration must also be given to the possible presence of corrosive or polluting substances.

709.512.2.1.4 Equipment should be located to avoid any foreseeable impact, be provided with local or general mechanical protection and have a degree of protection for external mechanical impact IK08 (see BS EN 62262).

Specific guidance for distribution boards and socket-outlets is given in section 8.9.

8.8 Wiring systems

709.521.1.5 The following wiring systems should **not** be used above a jetty, wharf, pier or pontoon:

▶ Cables suspended from or incorporating a support wire
▶ Non-sheathed cables in cable management systems e.g. conduit and trunking
▶ Cables with aluminium conductors
▶ Mineral insulated cables.

709.521.1.6 Conduit and ducting installations should have suitable apertures or holes and be fixed at an angle sloping away from the horizontal, sufficient to allow for drainage of moisture.

Cables should be selected and installed so that mechanical damage due to tidal and other movement of craft and other floating structures is prevented. To clarify this requirement, cables should be installed in such a manner that they are protected from damage due to:

▶ displacement by movement of craft or other structures
▶ friction, tension or crushing
▶ exposure to adverse temperatures.

See Figure 8.1 for a typical wiring arrangement for offshore pontoons.

At locations where cables are subject to flexing, e.g. bridge ramps, between movable jetties and pontoons, flexible cables should be used, such as:

▶ cross-linked insulated flexible cables harmonized type H07RN-F, H07BN4-F or H07RN8-F (insulated and sheathed), e.g. cables to Tables 14, 15, 16, 17 and 20 of BS 7919:2001
▶ thermosetting insulated flexible cables harmonized type H07Z-K, e.g. cables to BS 7211 Table 3b within flexible wiring systems.

Notes:

1 Cables should be installed in locations where they are protected from physical damage and, wherever practicable, out of water.

2 Many cable types including PVC insulated and sheathed cables are not suitable for continuous immersion in water. The suitability of the cable types should be checked with the manufacturers. Floating pontoons are usually manufactured with a service void in them, enclosed and accessible from above, to accommodate cables and water piping.

3 Fixed cables installed permanently under water at a depth of more than 4 m will normally need to be metal sheathed, e.g. lead. Alternatively, a cable that has been specially designed and manufactured for such locations could be used (refer to cable manufacturer for further guidance).

Fixed cables not permanently immersed or at a depth of less than 4 m should be armoured and incorporate extruded MDPE (medium density polyethylene) outer sheath. Note that cables used in this location will not have the same life expectancy as similar cables used on dry land.

Cable type H07 RN8 may be suitable up to 10 m depth of water. However, this type of cable does not offer any mechanical protection against, for example, boat anchors. Other types of cable (for example with mechanical protection) would need to be specially manufactured to the customer's requirements.

4 Due to the possibility of corrosion, the galvanised steel armouring of cables must not be used wholly or in part as a circuit protective conductor (cpc) on the floating section of marinas. A separate protective conductor should be used which, when sized in accordance with Regulation 543.1.2, can be common to several circuits if necessary. The armour must still, however, be connected to protective earth.

543.1.2

5 Protective bonding connections must be single-core PVC insulated to BS 6004 (HAR reference – H07V-R and H07Z-R), or BS 6007 (flexible type), or with an oversheath or further mechanical protection as applicable to the particular location.

6 Conductor colour coding should be in accordance with the requirements of BS 7671 Table 51. Terminations should be protected against corrosion either by the selection of suitable materials or covering with grease or water-resistant mastic or paint.

Table 51

7 Care should be exercised when installing cables to prevent damage from abrasion due to movement between pontoon sections, etc. Cables must be adequately fixed, protected and supported, and, if necessary, cable types suitable for the flexing movement must be used.

8 Where cables are installed at onshore locations, due consideration should be given to the routing, depth of lay and protection, especially where heavy traffic and point loads are experienced. Cables should normally be laid above the water table, or cable types suitable for continual immersion used. It is not usually practicable for buried cable duct systems to be made totally watertight. The watertight termination of ducts into drawpits and cable trenches below switchboards is also difficult to achieve.

▶ **Figure 8.1** Typical wiring arrangement from shore to pontoon

Notes:

1 Where the particular feeder pillars are in external locations they should be constructed of glass reinforced plastic (GRP), or have GRP housings. GRP is preferred to galvanised steel for protection against corrosion in such environments.

2 In order to counteract condensation within feeder pillar enclosures, low wattage 'anti-condensation' heaters should be installed.

3 All feeder pillar and distribution board doors should be fitted with locks to prevent unauthorised access, and have intermediate barriers to protect against accidental contact with live parts when the doors are open. The barriers should provide a degree of protection of at least IPXXB or IP2X.

8.9 Distribution boards, feeder pillars and socket-outlets

709.531.2   RCBOs must disconnect all poles.

Distribution boards and feeder pillars mounted outdoors should meet the degree of protection IP44 as a minimum. This will be adequate in sheltered waterways, but the IP code must be selected with reference to the degree of protection necessary for the particular location. The enclosure should be corrosion resistant and give protection against mechanical damage and ingress of dust and sand etc. Distribution boards and feeder

709.553.1.13   pillars supplying marina berths should be sited in the immediate vicinity of berths. When distribution boards and feeder pillars and their associated socket-outlets are mounted on floating installations or jetties, they should be fixed above the walkway and preferably not less than 1 m above the highest water level (see Figure 8.2). This height may be reduced to 300 mm if appropriate additional measures are taken to protect against the effects of splashing (IPX4), but care should be taken to avoid creating a low-level obstacle which may cause risk of tripping on the walkway. When mounted on fixed jetties they should be mounted not less than 1 m above the highest water level.

▼ **Figure 8.2**   Typical feeder pillar showing socket-outlet height for floating installation

709.553.1.10
709.553.1.11
709.531.2

709.553.1.8

709.553.1.13

Note: While there is a good argument for socket-outlets on feeder pillars and bollards to be mounted at a high level, they may be at risk from damage from the bows of boats which can accidentally overshoot the walkways during berthing. A lower mounting level of 300 mm minimum above the walkway can reduce this risk, but care should be taken to avoid creating a low level obstacle, which may trip the unwary.

709.553.1.9   Socket-outlets should be mounted as close as possible to the berth to be supplied
709.553.1.10   and should be installed in a distribution board or separate enclosures. A maximum of four socket-outlets may be grouped together in one enclosure. This is to minimise the hazard of long trailing flexes. A socket-outlet, either single-phase or polyphase, is only to supply a single pleasure craft or houseboat.

709.553.1.14   Socket-outlet protective conductors shall not be connected to a PME earthing facility.

709.553.1.8   Socket-outlets should be in accordance with BS EN 60309-2 and each outlet should be connected to the circuit protective conductor except where an onshore isolating transformer is used.

709.553.1.12   Generally, socket-outlets with a rating of 16 A should be provided and should have the following characteristics, irrespective of the measure of protection against electric shock:

Single-phase socket-outlets

Rated voltage: 230 V (colour blue)
Rated current: 16 A
Key position: 6 h
Number of poles: 2 plus protective conductor
Construction: IP44 (minimum)

Three-phase socket-outlets

Rated voltage: 400 V (colour red)
Rated current: 16 A
Key position: 6 h
Number of poles: 4 plus protective conductor
Construction: IP44 (minimum)

Note: Whilst 2P + E and 4P + E plugs and sockets are generally used, other configurations may be necessary as in the case of special security circuits indicating unauthorised use of particular socket-outlets on remote monitoring systems.

Where the pleasure craft demand is likely to exceed 16 A, provision should be made for outlets of suitable rating.

709.531   For automatic disconnection of supply, socket-outlets to supply pleasure craft and final circuits intended for fixed supplies to houseboats are to be protected individually by an RCD having the characteristics specified in Regulation 415.1.1. The devices will need to disconnect all poles including the neutral (see Figures 8.3a & b).

709.533   Socket-outlets and fixed connection supplies to houseboats are to be protected by individual overcurrent protective devices, which should comply with the requirements of Chapter 43 of BS 7671:2008 (see Figures 8.3a & b).

709.537.2.1.1   For a distribution cabinet there should be at least one means of isolation that will disconnect all live conductors including the neutral. If the device is used to isolate socket-outlets then it should isolate not more than four of these.

Preferably, socket-outlets or groups of single-phase socket-outlets intended for use on the same walkway or jetty should be connected to the same phase. However, individual sockets connected to separate phases of a supply should be located so that they cannot be reached simultaneously. If sockets on separate phases are grouped together on a pillar then a notice warning of the maximum voltage that exists between accessible parts should be provided.

Socket-outlets should include an interlock to prevent the insertion or removal of a plug while under load.

▼ Figure 8.3
General arrangements
for electricity supply to
pleasure craft

a Connection to mains supply with single-phase socket-outlet

b Connection to mains supply with three-phase socket-outlet

A notice of durable material giving instructions for connection of a pleasure craft to the marina supply is recommended to be placed where practicable adjacent to each group of socket-outlets, bearing indelible, weatherproofed and easily legible characters.

Alternatively, the notice should be placed in a prominent position or issued to each berth holder. The notice should contain the text of Figure 8.4.

▼ **Figure 8.4**
Instruction notice for connection of supply (Figure 709.3 of BS 7671:2008)

INSTRUCTIONS FOR ELECTRICITY SUPPLY

BERTHING INSTRUCTIONS FOR CONNECTION TO SHORE SUPPLY

This marina provides power for use on your pleasure craft with a direct connection to the shore supply which is connected to earth. Unless you have an isolating transformer fitted on board to isolate the electrical system on your craft from the shore supply system, corrosion through electrolysis could damage your craft or surrounding craft.

ON ARRIVAL

1   Ensure the supply is switched off and disconnect all current-using equipment on the craft, before inserting the craft plug. Connect the flexible cable **firstly** at the pleasure-craft inlet socket and **then** at the marina socket-outlet.

2   The supply at this berth is *V, *Hz. The socket-outlet will accommodate a standard marina plug colour * (technically described as BS EN 60309 2, position 6 h).

3   For safety reasons, your craft must not be connected to any other socket-outlet than that allocated to you and the internal wiring on your craft must comply with the appropriate standards.

4   Every effort must be made to prevent the connecting flexible cable from falling into the water if it should become disengaged. For this purpose, securing hooks are provided alongside socket-outlets for anchorage at a loop of tie cord.

5   For safety reasons, only one pleasure craft connecting cable supplying one pleasure craft may be connected to any one socket-outlet.

6   The connecting flexible cable must be in one length, without signs of damage , and not contain joints or other means to increase its length.

7   The entry of moisture and salt into the pleasure craft inlet socket may cause a hazard. Examine carefully and clean the plug and socket before connecting the supply.

8   It is dangerous to attempt repairs or alterations. If any difficulty arises, contact the marina management.

BEFORE LEAVING

1   Ensure that the supply is switched off and disconnect all current-using equipment on the craft, before the connecting cable is disconnected and any tie cord loops are unhooked.

2   The connecting flexible cable should be disconnected **firstly** from the marina socket-outlet and **then** from the pleasure craft inlet socket. Any cover that may be provided to protect the inlet from weather should be securely replaced. The connecting flexible cable should be coiled up and stored in a dry location where it will not be damaged.

---

*   Appropriate figures and colours to be inserted:
nominally 230 V 50 Hz blue – single-phase, and
nominally 400 V 50 Hz red – three-phase.

8.10 General notes

8.10.1 Pontoon amenity lighting

It is important that the routes of pontoons and their termination points are clearly delineated.

The lighting may be controlled by either automatic photoelectric cells or time switches, the former being preferred as they sense poor conditions caused by fog, etc. when natural light is waning.

709.512   Luminaires should be of rugged and watertight construction and should preferably be mounted at low level with the light source facing the walkway, not omnidirectional.

8.10.2 Navigation lighting

The local waterway authority should be consulted in order that all necessary and suitably coloured navigation lighting is provided. The light sources should have an extended life expectancy. Photoelectric cell control is preferred to time switches.

8.10.3 Fuelling stations

The relevant local authority should be consulted in order to ensure that the completed installation complies with its requirements. Where applicable, special emergency control facilities should be established onshore. Fuel hoses are required to be non-conducting. Ship/shore bonding cables are not to be used – see *The International Safety Guide for Oil Tankers and Terminals*, 4th Edition.

Electrical equipment in the proximity of fuelling stations should comply with APEA/IP *Guidance for the Design, Construction, Modification, Maintenance and Decommissioning of Filling Stations* (March 2005).

8.10.4 Metering systems

Metering systems are outside the scope of this Guidance Note and must be agreed between the designer and the marina owner to provide all necessary electricity consumption information for accurate billing. The meters may be required to be installed locally in the feeder pillars for local direct reading, or may be part of a site-wide data network system. The metering system must be fit for the installation and type of use. Functional and safety earthing must be adequate.

Check metering for the various main sections of the distribution system may be required in order that the marina operator can use this data in establishing tariffs for the resale of electricity. Such equipment must be installed within the main switchgear and feeder pillars, and must be of adequate rating and quality for the duty required. The increased use of items of electrical equipment exhibiting low power factor characteristics, e.g. dehumidifiers, refrigerators, battery chargers, etc., requires that electricity metering should record suitable data to ensure the marina operator does not suffer a loss of revenue. (This particularly applies when kVAh metering is installed by the electricity distributor.)

8.10.5 Location of equipment

Due consideration should be given to the location of items of equipment so that they are, as far as practicable, not vulnerable to damage either on- or offshore at the marina.

In the case of onshore areas there will be the need for clear vehicular movement including large mobile boat hoists, transit lorries and cars, etc. The location therefore of feeder pillars and lighting columns requires special attention.

For marina areas, the lighting columns and power supply feeder pillars should be so positioned that the risk of contact with luggage trolleys etc. and such items as the bowsprit of craft is, as far as practicable, reduced to a minimum. This is particularly important where lighting and power supply equipment has moulded enclosures which are unable to withstand such mechanical forces and impact and may be damaged.

Site investigations should be carried out at an early stage to determine maximum wave heights which can be experienced. This is of particular importance at exposed coastal sites. Where marinas have breakwater type pontoons, it is likely that under certain conditions waves will pass over the structure.

8.10.6 Routine maintenance and testing

Initial inspection and testing of all electrical systems should be carried out on completion of the installation, in accordance with the requirements of Part 6 of BS 7671 and the recommendations of IET Guidance Note 3. A periodic inspection and test of all electrical systems should be carried out annually and the necessary maintenance work implemented. If the site is considered to be exposed, or operational experience shows problems (i.e. misuse), the inspection frequency should be increased to cater for the particular conditions experienced.

All RCDs should be tested regularly by operating the integral test button and periodically by a proprietary instrument to ensure they conform with the parameters of their relevant product standards, e.g. BS 4293, BS EN 61008.

All tests should be tabulated for record purposes and the necessary forms required by Part 6 of BS 7671 must be provided by the contractor or persons carrying out the inspection and tests to the person ordering the work.

Medical locations 9

9.1 Introduction and scope

Introduction to Amendment No 1 to the 17th Edition

Sect 710 Section 710 'Medical locations', is a new section introduced into Part 7 (Special Installations or Locations) of BS 7671: 2008 by Amendment No 1:2011.

The section is based on the published standard IEC 60364-7-710: 2002 as modified by draft CENELEC standard prHD 60364-7-710, plus additional regulations to satisfy the UK Healthcare requirements.

701.1 The section applies to electrical installations in medical locations to ensure safety of patients and medical staff. These requirements, in the main, refer to hospitals, private clinics, medical and dental practices, healthcare centres and dedicated medical rooms in the workplace.

The requirements of this section do not apply to medical electrical equipment.

The section also applies to electrical installations in locations designed for medical research.

The use of medical electrical equipment can be split into three categories; some examples are listed below:

a **Life support:** infusion pumps, ventilators and dialysis machines etc
b **Diagnostic:** X-ray machines, CT scanners, magnetic resonance imagers, blood pressure monitors, electroencephalograph (EEG) and electrocardiograph (ECG) equipment
c **Treatment:** surgical diathermy and defibrillators etc.

710.56 Patients undergoing acute care require enhanced reliability and safety of the electrical installation in hospitals to ensure the security of supplies and minimise the incidence of electric shock.

Notes:
1 Supporting information on the electrical installations of medical locations can be found in *Health Technical Memorandum (HTM) 06-01 (Part A)*, published by the Department of Health.
2 For medical electrical equipment and medical electrical systems, refer to the BS EN 60601 series.

9.2 The risks

In medical locations stringent measures are necessary to ensure the safety of patients likely to be subjected to the application of medical electrical equipment.

Shock hazards due to bodily contact with the 50 Hz mains supply are well known and documented. Currents of the order of 10 mA passing through the human body can result in muscular paralysis followed by respiratory paralysis depending on skin resistances, type of contact, environmental conditions and duration. Eventual ventricular fibrillation can occur at currents just exceeding 20 mA. These findings are listed in IEC/TR2 60479-1 *Effects of current on human beings and livestock – General aspects*.

The natural protection of the human body is considerably reduced when certain clinical procedures are being performed on it. Patients under treatment may have their skin resistance broken or their defensive capacity either reduced by medication or nullified while anaesthetised. These conditions increase the possible consequences of a shock under fault conditions.

In patient environments where intracardiac procedures[1] are undertaken, the electrical safety requirements are even stricter, in order to protect the patient against 'microshock'. Patient leakage currents from applied parts introduced directly to the heart can interfere with cardiac function at current levels which would be considered safe under other circumstances.

Patient leakage current which can flow into an earthed patient is normally greatest when the equipment earth is disconnected. A limit is set to the amount of leakage current which can flow in the patient circuit when the protective earth conductor is disconnected. Patient leakage currents[2] of the order of 10 µA have a probability of 0.2 per cent for causing ventricular fibrillation or pump failure when applied through a small area of the heart. At 50 µA (microshock), the probability of ventricular fibrillation increases to the order of 1 per cent (refer to BS EN 60601-1).

Additional to the consideration of risk from electric shock, some electrical equipment (life-support equipment, surgical equipment) performs such vital functions that loss of supply would pose an unacceptable risk to patients. Medical locations where such

---

1 A procedure whereby an electrical conductor is placed within the heart of a patient or is likely to come into contact with the heart, such conductor being accessible outside the patient's body. In this context, an electrical conductor includes insulated wires such as cardiac pacing electrodes or intracardiac ECG electrodes, or insulated tubes filled with conducting fluids (catheter).
2 'Patient leakage current': current flowing from a medical electrical equipment applied part via the patient to earth.

equipment is used require secure supplies. This has implications not only for the provision of safety (emergency) power supplies, but can also render some conventional protective measures unsuitable. Hence, for example, when protecting circuits supplying critical medical equipment, restrictions are stipulated on the use of RCDs.

9.3 Definitions

9.3.1 Medical location

Location intended for purposes of diagnosis, treatment including cosmetic treatment, monitoring and care of patients.

9.3.2 Patient

Living being (person or animal) undergoing a medical, surgical or dental procedure. A person under treatment for cosmetic purposes may be considered a patient.

9.3.3 Medical electrical equipment (ME equipment)

Electrical equipment having an applied part or transferring energy to or from the patient or detecting such energy transfer to or from the patient and which is

a provided with not more than one connection to a particular supply mains, and
b intended by the manufacturer to be used
  ▶ in the diagnosis, treatment or monitoring of a patient, or
  ▶ for compensation or alleviation of disease, injury or disability.

Note: ME equipment includes those accessories as defined by the manufacturer that are necessary to enable the normal use of the ME equipment.

9.3.4 Applied part

Part of medical electrical equipment that in normal use necessarily comes into physical contact with the patient for ME equipment or an ME system to perform its function.

9.3.5 Group 0

Medical location where no applied parts are intended to be used and where discontinuity (failure) of the supply cannot cause danger to life.

9.3.6 Group 1

Medical location where discontinuity of the electrical supply does not represent a threat to the safety of the patient and applied parts are intended to be used:

i externally
ii invasively to any part of the body, except where Group 2 applies.

9.3.7 Group 2

Medical location where applied parts are intended to be used, and where discontinuity (failure) of the supply can cause danger to life, in applications such as:

i intracardiac procedures
ii vital treatment and surgical operations.

9.3.8 Medical electrical system (ME system)

Combination, as specified by the manufacturer, of items of equipment, at least one of which is medical electrical equipment to be interconnected by functional connection or by use of a multiple socket-outlet.

Note: The system includes those accessories which are needed for operating the system and are specified by the manufacturer.

9.3.9 Patient environment

Any volume in which intentional or unintentional contact can occur between a patient and parts of the medical electrical equipment or medical electrical system or between a patient and other persons touching parts of the medical electrical equipment or medical electrical system (for illustration see Figure 9.1).

▼ **Figure 9.1**
Patient environment
(BS EN 60601-1:2006)

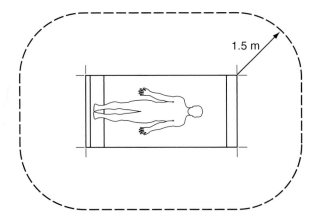

Note: The dimensions in the figure show the minimum extent of the patient environment in a free surrounding. It applies when the patient's position is pre-determined; if not, all possible patient positions should be considered.

9.3.10 Medical IT system

IT electrical system fulfilling specific requirements for medical applications.

Note: These supplies are also known as isolated power supply systems (IPS).

9.4 Assessment of general characteristics

710.3 In order to determine the classification and group number of a medical location, it is necessary that the relevant medical staff indicate which medical procedures will take place within the location. Based on the intended use, the appropriate classification for the location can be determined.

Notes:

1 Guidance on the allocation of a group number and classification of safety services for medical locations is shown in Table 9.1.

2 The possibility that certain medical locations could be used for different purposes may require a higher group allocation.

Medical locations[c]	Group			Classification (see 9.18.1)	
	0	1	2	≤0.5 s	>0.5 s ≤15 s
1 Massage room	X	X			X
2 Bedrooms		X			X
3 Delivery room		X		X[a]	X
4 ECG, EEG, EHG room		X			X
5 Endoscopic room		X[b]		X	X[b]
6 Examination or treatment room		X		X	X
7 Urology room		X[b]		X	X[b]
8 Radiological diagnostic and therapy room		X	X	X	X
9 Hydrotherapy room		X			X
10 Physiotherapy room		X			X
11 Anaesthetic area			X	X[a]	X
12 Operating theatre			X	X[a]	X
13 Operating preparation room			X	X[a]	X
14 Operating plaster room			X	X[a]	X
15 Operating recovery room			X	X[a]	X
16 Heart catheterization room			X	X[a]	X
17 Intensive care room			X	X[a]	X
18 Angiographic examination room			X	X[a]	X
19 Haemodialysis room			X		X
20 Magnetic resonance imaging (MRI) room	X	X		X	X
21 Nuclear medicine		X			X
22 Premature baby room			X	X[a]	X
23 Intermediate Care Unit (IMCU)			X	X	X

Notes on Table 9.1:

a Luminaires and life-support ME equipment which needs power supply within 0.5 s or less.

b Not being an operating theatre.

c This list is provided as a guide only and agreement has to be reached with local management.

9.5 Type of system earthing

710.312.2 PEN conductors are not to be used in medical locations and medical buildings downstream of the main distribution board.

Note: In Great Britain, Regulation 8(4) of the Electricity Safety, Quality and Continuity Regulations 2002 prohibits the use of PEN conductors in consumers' installations.

9.6 Supplies

710.313.1 In medical locations, the distribution system should be designed and installed to facilitate the automatic changeover from the main distribution network to the electrical safety source feeding essential loads.

710.512.1.2 9.6.1 Power supply for medical locations of Group 2

In the event of a first fault to earth, a total loss of supply in Group 2 locations must be prevented.

9.7 Protection against electric shock

710.410.3.5 Protective measures providing basic protection (protection against direct contact) only, by obstacles and placing out of reach, are not permitted.

710.410.3.6 The protective measures of non-conducting location, earth-free local equipotential bonding and electrical separation for the supply of more than one item of current-using equipment, are not permitted

Note: Only protection by insulation of live parts, by barriers or enclosures or by the use of Class II equipment are permitted.

9.8 Requirements for fault protection (protection against indirect contact)

9.8.1 Automatic disconnection in case of a fault

710.411.3.2.1 Care shall be taken to ensure that simultaneous use of many items of equipment connected to the same circuit cannot cause unwanted tripping of the residual current protective device (RCD).

710.411.3.2.1 In medical locations of Group 1 and Group 2, where RCDs are required, only type A or type B shall be selected, depending on the possible fault current arising. Type AC RCDs shall not be used.

Note: Type 'A' ensures tripping for residual sinusoidal alternating currents and residual pulsating direct currents, whether suddenly applied or slowly rising. Type 'B' ensures tripping for residual sinusoidal alternating currents and residual pulsating direct currents and smooth direct currents.

710.411.3.2.5 In medical locations of Group 1 and Group 2, the following shall apply:

i For IT, TN and TT systems, the voltage presented between simultaneously accessible exposed-conductive-parts and/or extraneous-conductive-parts shall not exceed 25 V a.c. or 60 V d.c.

ii For TN and TT systems, the requirements of Table 9.2 shall apply.

▼ **Table 9.2** Maximum disconnection times for TN and TT systems in Groups 1 and 2 locations

System	25 V < U_0 ≤ 50 V (seconds)		50 V < U_0 ≤ 120 V (seconds)		120 V < U_0 ≤ 230 V (seconds)		230 V < U_0 ≤ 400 V (seconds)		U_0 > 400 V (seconds)	
	a.c.	d.c.	a.c.	d.c.	a.c.	d.c.	a.c.	d.c.	a.c.	d.c.
TN	5	5	0.3	2	0.3	0.5	0.05	0.06	0.02	0.02
TT	5	5	0.15	0.2	0.05	0.1	0.02	0.06	0.02	0.02

Note: In TN systems, a value of 25 V a.c. or 60 V d.c. **may** be met with protective equipotential bonding, complying with the disconnection time in accordance with Table 9.2 (refer also to the notes associated with 9.12.1).

9.8.2 Additional protection

710.411.3.3 Where a medical IT system is used, additional protection by means of an RCD shall not be used.

710.411.4 ### 9.8.3 TN system

In final circuits of Group 1 rated up to 63 A, RCDs with a rated residual operating current not exceeding 30 mA and meeting the requirements of Regulation 415.1.1 shall be used.

In TN-S systems, the insulation level of all live conductors shall be monitored.

Note 1: Monitoring of TN-S systems is required to detect any Neutral to Earth faults implying a TN-C system, which is not permitted (refer to 9.5). The scenario can lead to excessive currents flowing in the earthed enclosures (exposed-conductive-parts) of medical systems inducing higher touch voltages within Group 2 locations beyond the limits allowed. High circulating currents in earthed enclosures can also cause malfunction of sensitive electronic ME equipment.

Note 2: An indirect method of monitoring the insulation of the neutral conductor is to measure the leakage current in the PE conductor using a Residual Current Monitor (RCM). The latter can be set up to supervise several zones of the medical location and pinpoint a fault within a respective zone where an alarm is initiated.

In medical locations of Group 2 (except for the medical IT system), protection by automatic disconnection of supply by means of RCDs with a rated residual operating current not exceeding 30 mA and meeting the requirements of Regulation 415.1.1 shall only be used on the following circuits:

i circuits for the supply of movements of fixed operating tables, or
ii circuits for X-ray units, or
iii circuits for large equipment with a rated power greater than 5 kVA.

Note: The requirement in (ii) above is mainly applicable to mobile X-ray units brought into Group 2 locations.

710.411.5 ### 9.8.4 TT system

In medical locations of Group 1 and Group 2, RCDs shall be used as disconnection devices and the requirements of 9.8.3 shall apply.

710.411.6 **9.8.5 IT system**

710.411.6.3.1 In Group 2 medical locations, an IT system shall be used for final circuits supplying ME equipment and systems intended for life support, surgical applications and other electrical equipment located or that may be moved into the 'patient environment', excluding equipment listed in 9.8.3.

For each group of rooms serving the same function, at least one IT system is necessary. The IT system shall be equipped with an insulation monitoring device (IMD) in accordance with BS EN 61557-8:2007, with the following additional specific requirements:

i   a.c. internal impedance shall be ≥ 100 kΩ
ii internal resistance shall be ≥ 250 kΩ
iii test voltage shall be ≤ 25 V d.c.
iv injected current, even under fault conditions, shall be ≤ 1 mA peak
v   indication shall take place at the latest when the insulation resistance has decreased to 50 kΩ. If the response value is adjustable, the lowest decreased possible setpoint value shall be ≥ 50 kΩ. A test device shall be provided
vi response and alarm-off time shall be ≤ 5 s.

Note: An indication is recommended if the protective earth (PE) or wiring connection of the IMD is lost.

▼ **Figure 9.2**   Typical theatre layout of medical IT system with insulation monitoring

For each medical IT system, an acoustic and visual alarm system incorporating the following components shall be arranged at a suitable place so that it can be permanently monitored (audible and visual signals) by the medical staff and, furthermore, is forwarded to the technical staff:

vii a green signal lamp to indicate normal operation
viii a yellow signal lamp which lights when the minimum value set for the insulation resistance is reached. It shall not be possible for this light to be cancelled or disconnected

ix   an audible alarm which sounds when the minimum value set for the insulation resistance is reached. This audible alarm may be silenced

x   the yellow signal shall go out on removal of the fault and when the normal condition is restored.

Documentation shall be easily readable in the medical location and it shall include:

xi   the meaning of each type of signal, and

xii the procedure to be followed in case of an alarm at first fault.

Note: For illustration of a typical theatre layout refer to Figure 9.2.

710.411.6.3.2   Monitoring of overload and high temperature for the IT transformer is required.

710.411.6.3.3   Fault location systems which localize insulation faults in any part of the medical IT system may also be installed in addition to an insulation monitoring device. The insulation fault location system shall be in accordance with BS EN 61557-9.

710.512.1.1   ## 9.9   Transformers for IT systems

Transformers shall be in accordance with BS EN 61558-2-15, installed in close proximity to the medical location and with the following additional requirements:

i   The leakage current of the output winding to earth and the leakage current of the enclosure, when measured in no-load condition and the transformer supplied at rated voltage and rated frequency, shall not exceed 0.5 mA

ii   At least one single-phase transformer per room or functional group of rooms shall be used to form the IT systems for portable and fixed equipment and the rated output shall be not less than 0.5 kVA and shall not exceed 10 kVA. Where several transformers are needed to supply equipment in one room, they shall not be connected in parallel

iii If the supply of three-phase loads via an IT system is also required, a separate three-phase transformer shall be provided for this purpose.

For monitoring see 9.8.5.

Capacitors shall not be used in transformers for medical IT systems.

710.411.7   ## 9.10 Functional extra-low voltage (FELV)

In medical locations, functional extra-low voltage (FELV) is not permitted as a method of protection against electric shock.

710.414.1   ## 9.11 Protection by SELV or PELV

When using SELV and/or PELV circuits in medical locations of Group 1 and Group 2, the nominal voltage applied to current-using equipment shall not exceed 25 V rms a.c. or 60 V ripple-free d.c. Protection by basic insulation of live parts as required by Regulation 416.1 or by barriers or enclosures as required by Regulation 416.2, shall be provided.

710.414.4.1   In medical locations of Group 2, where PELV is used, exposed-conductive-parts of equipment, e.g. operating theatre luminaires, shall be connected to the circuit protective conductor.

9.12 Supplementary equipotential bonding (additional protection)

710.415.2.1 In each medical location of Group 1 and Group 2, supplementary equipotential bonding shall be installed and the supplementary bonding conductors shall be connected to the equipotential bonding busbar for the purpose of equalizing potential differences between the following parts, which are located or that may be moved into the 'patient environment':

i   protective conductors
ii   extraneous-conductive-parts
iii screening against electrical interference fields, if installed
iv   connection to conductive floor grids, if installed
v   metal screen of isolating transformers, via the shortest route to the earthing conductor.

Supplementary equipotential bonding connection points for the connection of medical electrical equipment shall be provided in each medical location, as follows:

vi   Group 1: one per patient location
vii Group 2: one per medical IT socket-outlet.

Note: Fixed conductive non-electrical patient supports such as operating theatre tables, physiotherapy couches and dental chairs should be connected to the equipotential bonding conductor unless they are intended to be isolated from earth.

710.415.2.2 ### 9.12.1 Resistance of protective conductors in Group 1 locations

In medical locations of Group 1, the resistance of the protective conductors, including the resistance of the connections, between the terminals for the protective conductor of socket-outlets and of fixed equipment or any extraneous-conductive-parts and the equipotential bonding busbar shall not exceed 0.7 ohm.

Notes:

1   TN systems in Group 1 locations are protected by RCDs designed to operate within 0.3 s.
2   The designer should check that the touch voltage cannot exceed 25 V a.c. or 60 V ripple-free d.c. under single-fault conditions.

9.12.2 Resistance of protective conductors in Group 2 locations

710.415.2.2 In medical locations of Group 2, the resistance of the protective conductors, including the resistance of the connections, between the terminals for the protective conductor of socket-outlets and of fixed equipment or any extraneous-conductive-parts and the equipotential bonding busbar shall not exceed 0.2 ohm.

Notes:

1   TN systems in Group 2 locations are protected by RCDs designed to operate within 0.3 s.
2   The 0.2 ohm will serve to limit any potential rise due to the effect of leakage current from ME equipment. The voltage between the hard-wired system and the Earth Reference Bar (ERB) should not exceed 20 mV. A further 30 mV is allowed between the exposed-conductive-parts of the ME equipment and the supply cable, making a maximum of 50 mV. This would contribute to keeping the microshock currents within tolerable limits.

3 The ERB is located within the vicinity of the Group 2 location to allow short lengths of bonding conductors to be connected (for example 6 mm^2) where the resistive values remain within the 0.2 ohm. The additional protective earth bar is located some distance away within the vicinity of the distribution board of the IT transformer panel to allow the connection between the ERB and this panel by a larger size protective earthing conductor (for example 10 mm^2). Refer to Figure 9.2.

710.415.2.3 ### 9.12.3 Specification of the equipotential bonding busbar

The equipotential bonding busbar shall be located in or near the medical location. Connections shall be so arranged that they are accessible, labelled, clearly visible and can easily be disconnected individually.

710.51.1 ## 9.13 Distribution boards

Distribution boards for Group 2 should be installed in close proximity to the Group 2 medical locations and clearly identified.

710.512.2.1 ## 9.14 Explosion risk

Electrical devices, e.g. socket-outlets and switches, installed below any medical-gas outlets for oxidizing or flammable gases shall be located at a distance of at least 0.2 m from the outlet (centre to centre), so as to minimise the risk of ignition of flammable gases.

Note: Requirements for medical electrical equipment for use in conjunction with flammable gases and vapours are contained in BS EN 60601-1.

710.52 ## 9.15 Wiring systems in Group 2 locations

Any wiring system within Group 2 medical locations shall be exclusive for the use of equipment and fittings within those locations.

Note: It is recommended that radial wiring patterns are used to avoid "earth-loops" that may cause electromagnetic interference.

710.531.1
710.52 ### 9.15.1 Protection of wiring systems in medical locations of Group 2

Overload current protection shall not be used in either the primary or secondary circuit of the transformer of a medical IT system.

Overcurrent protection against short-circuit and overload current is required for each final circuit.

Note: Overcurrent protective devices (e.g. fuses) may be used in the primary circuit of the transformer for short-circuit protection only.

710.559 ## 9.16 Lighting circuits

In medical locations of Group 1 and Group 2, at least two different sources of supply shall be provided. One of the two sources shall be connected to the electrical supply system for safety services.

9.16.1 Safety lighting

710.560.9.1

In the event of mains power failure, the changeover period to the safety services source shall not exceed 15 s. The necessary minimum illuminance shall be provided for the following:

i emergency lighting and exit signs to BS 5266

ii locations for switchgear and controlgear for emergency generating sets and for main distribution boards of the normal power supply and for power supply for safety services

iii rooms in which essential services are intended. In each such room at least one luminaire shall be supplied from the power source for safety services

iv locations of central fire alarm and monitoring systems

v rooms of Group 1 medical locations. In each such room at least one luminaire shall be supplied from the power supply source for safety services

vi rooms of Group 2 medical locations. A minimum of 90 per cent of the lighting shall be supplied from the power source for safety services.

The luminaires of the escape routes shall be arranged on alternate circuits.

9.17 Socket-outlets

9.17.1 Circuits protected by RCDs

710.531.2.4

For each circuit protected by an RCD with a rated residual operating current not exceeding 30 mA and meeting the requirements of Regulation 415.1.1, consideration shall be given to reduce the possibility of unwanted tripping of the RCD due to excessive protective conductor currents produced by equipment in normal operation.

9.17.2 Circuits in the medical IT system for medical locations of Group 2

710.553.1

Socket-outlets intended to supply medical electrical equipment shall be unswitched.

At each patient's place of treatment, e.g. bedheads, the configuration of socket-outlets shall be as follows:

▶ each socket-outlet supplied by an individually protected circuit; or

▶ several socket-outlets separately supplied by a minimum of two circuits.

Socket-outlets used on medical IT systems shall be coloured blue and clearly and permanently marked "Medical Equipment Only".

Note 1: Refer also to *Health Technical Memorandum (HTM) 06-01 Part (A)* published by the Department of Health.

Note 2: The two circuits are normally fed from the same IT transformer where each circuit is individually protected.

9.18 Safety services

710.56

In medical locations a power supply for safety services is required, which, in accordance with BS 7671, will energize the installations needed for continuous operation in case of failure of the general power system, for a defined period with a preset switch-over time.

The safety power supply system shall automatically take over if the voltage of one or more incoming live conductors of the main distribution board of the building with the main power supply has dropped for more than 0.5 s and by more than 10 per cent in regard to the nominal voltage.

710.560.4 ### 9.18.1 Classification of safety services for medical locations

▼ **Table 9.3**
Classification of safety services necessary for medical locations

Classification	Changeover time (s)	Description
No-break	0	Automatic supply available with no break
Very short break	0.15	Automatic supply available within 0.15 s
Short break	0.5	Automatic supply available within 0.5 s
Medium break	15	Automatic supply available within 15 s
Long break	>15	Automatic supply available in more than 15 s

Notes:

1   Generally it is unnecessary to provide a no-break power supply for medical electrical equipment. However, certain microprocessor controlled equipment may require such a supply.
2   Safety services provided for locations having differing classifications should meet that classification which gives the highest security of supply. Refer to Table 9.1 for guidance on the association of classification of safety services with medical locations.

710.560.5.5 ### 9.18.2 General requirements for safety power supply sources of Groups 1 and 2

Primary cells are not allowed as safety power sources.

An additional main incoming power supply, from the general power supply, is not regarded as a source of the safety power supply.

The availability (readiness for service) of safety power supply sources shall be monitored and indicated at a suitable location.

710.560.5.6 ### 9.18.3 Failure of the general power supply source

In case of a failure of the general power supply source, the power supply for safety services shall be energized to feed the equipment stated in 9.18.5.1, 9.18.5.2 and 9.18.5.3 with electrical energy for a defined period of time and within a predetermined changeover period.

710.560.5.7 ### 9.18.4 Socket-outlets

Where socket-outlets are supplied from the safety power supply source they shall be readily identifable according to their safety services classification.

9.18.5 Electrical sources for safety services

710.560.6.1.1 #### 9.18.5.1 Power supply sources with a changeover period ≤ 0.5 s

In the event of a voltage failure on one or more line conductors at the distribution board, a safety power supply source shall be used and be capable of providing power for a period of at least 3 h for:

- luminaires of operating theatre tables
- medical electrical equipment containing light sources being essential for the application of the equipment, e.g. endoscopes, including associated essential equipment, e.g. monitors
- life-supporting medical electrical equipment.

The normal power supply shall be restored within a changeover period not exceeding 0.5 s.

Note: The duration of 3 h may be reduced to 1 h if a power source meeting the requirements of 9.18.5.2 is installed.

710.560.6.1.2 **9.18.5.2 Power supply sources with a changeover period ≤ 15 s**

Equipment meeting the requirements of 9.16.1 and 9.18.7 shall be connected within 15 s to a safety power supply source capable of maintaining it for a minimum period of 24 h, when the voltage of one or more live conductors at the main distribution board for the safety services has decreased by more than 10 per cent of the nominal value of supply voltage and of a duration greater than 3 s.

710.560.6.1.3 **9.18.5.3 Power supply sources with a changeover period >15 s**

Equipment, other than that covered by sections 9.18.5.1 and 9.18.5.2, which is required for the maintenance of hospital services, may be connected either automatically or manually to a safety power supply source capable of maintaining it for a minimum period of 24 h. This equipment may include, for example:

- sterilisation equipment
- technical building installations, in particular air conditioning, heating and ventilation systems, building services and waste disposal systems
- cooling equipment
- catering equipment
- storage battery chargers.

710.560.7 **9.18.6 Circuits of safety services**

The circuit which connects the power supply source for safety services to the main distribution board should be considered a safety circuit.

710.560.11 **9.18.7 Other services**

Other services which may require a safety service supply with a changeover period not exceeding 15 s include, for example, the following:

- selected lifts for firefighters
- ventilation systems for smoke extraction
- paging systems
- medical electrical equipment used in Group 2 medical locations which serves for surgical or other procedures of vital importance. Such equipment will be defined by responsible staff
- electrical equipment of medical gas supply including compressed air, vacuum supply and narcosis (anaesthetics) exhaustion as well as their monitoring devices
- fire detection and fire alarms to BS 5839
- fire extinguishing systems.

9.19 Inspection and testing

9.19.1 Initial verification

710.61 The dates and results of each verification shall be recorded.

Chap 61 The tests specified below under items **i** to **iii** in addition to the requirements of Chapter 61 and HTM 06-01 (Part A) shall be carried out, both prior to commissioning and after alterations or repairs and before re-commissioning:

i Complete functional tests of the insulation monitoring devices (IMDs) associated with the medical IT system including insulation failure, transformer high temperature, overload, discontinuity and the acoustic/visual alarms linked to them
ii Measurements of leakage current of the output circuit and of the enclosure of the medical IT transformers in no-load condition; refer to 9.9
iii Measurements to verify that the resistance of the supplementary equipotential bonding is within the limits stipulated by 9.12.1 and 9.12.2.

710.62 ### 9.19.2 Periodic inspection and testing

Notes:
1 Periodic inspection and testing should be carried out in accordance with *Health Technical Memorandum (HTM) 06-01 (Part B)* and local Health Authority requirements, if any.
Chap 62 **2** In addition to the requirements of Chapter 62, the following procedures are recommended at the given intervals:
i Annually – Complete functional tests of the insulation monitoring devices (IMDs) associated with the medical IT system including insulation failure, transformer high temperature, overload, discontinuity and the acoustic/visual alarms linked to them
ii Every 3 years – Measurements of leakage current of the output circuit and of the enclosure of the medical IT transformers in no-load condition; refer to 9.9
iii Annually – Measurements to verify that the resistance of the supplementary equipotential bonding is within the limits stipulated by 9.12.1 and 9.12.2.

9.20 Recommendations from the Department of Health

The Department of Health provides medical, building and engineering guidance through its *Health Technical Memoranda (HTM)* series of documents. Five documents are specifically related to electrical engineering issues.

The electrical series of documents are supported by the general policies and procedures embraced in HTM 00. These documents supersede HTM 2007, 2011, 2014, 2020 and 2021.

The current titles are:

HTM 00	Policies and Procedures (2006)
HTM 06-01 Part A	Electrical Services supply and distribution; Design considerations (2007)
HTM 06-01 Part B	Electrical Services supply and distribution; Operational management (2007)
HTM 06-02	Electrical Safety Guidance for low voltage systems (2006)
HTM 06-02	Electrical Safety Handbook
HTM 06-03	Electrical Safety Guidance for high voltage systems (2006)

HTM 06-01 Part A provides guidance for all works on the fixed wiring and integral electrical equipment used for electrical services within healthcare premises. The document should be used for all forms of electrical design work ranging from a new greenfield site to modifying an existing final circuit.

HTM 06-01 Part B addresses the operational management and maintenance of the electrical services supply distribution within a healthcare facility.

HTM 06-02, together with its associated *Safety Guidance Handbook*, gives operational guidance on electrical safety requirements for low voltage systems in healthcare premises.

HTM 06-03 gives operational guidance on electrical safety requirements for high voltage systems in healthcare premises.

Solar photovoltaic (PV) power supply systems

10

10.1 Introduction

Sect 712

More and more solar photovoltaic (PV) systems are being installed with the aim of cutting carbon dioxide emission and reducing electricity bills. Section 712 of BS 7671 has specific requirements for the installation of PV systems.

Note: A *Guide to the Installation of PV systems* (2nd edition) by the DTI is available from the Department for Business, Innovation and Skills (BIS): see *www.bre. co.uk/filelibrary/pdf/rpts/Guide_to_the_installation_of_PV_systems_2nd_ Edition.pdf*. This DTI guide is currently being rewritten as *The Installer Guide to Photovoltaic Installations* 978-0-901452-48-1 – under the auspices of the MCS scheme.

Amendment No 1 to the 17th Edition

There are no significant changes introduced in Section 712 by Amendment No 1 to the 17th Edition.

Introduction to the 17th Edition

This is a new section that has been introduced into the 17th Edition. There are an increasing number of solar power supplies that are being installed in all kinds of premises, with their own particular risks for installers, users and maintenance personnel.

The section includes requirements for accessibility, external influences, routing of protective conductors, selection and erection of cables to avoid the risk of lightning strike, short-circuit and earth faults, overcurrent protection, compliance with standards, protection against electromagnetic interference, and devices for isolation.

10.2 Scope

712.1

The particular requirements of Section 712 apply to the electrical installations of PV power supply systems including systems with a.c. modules.

Note: Requirements for PV power supply systems which are intended for stand-alone operation are under consideration.

10.3 The Electricity Safety, Quality and Continuity Regulations 2002

The solar photovoltaic (PV) power supply systems described in this chapter are required to meet the requirements of the Electricity Safety, Quality and Continuity Regulations 2002, as they are embedded generators.

Where the output does not exceed 16 A per phase they are by definition small-scale embedded generators (SSEG) and are exempted from certain of the requirements of Regulation 22 provided that:

i   the equipment should be type tested and approved by a recognised body,

ii the consumer's installation must comply with the requirements of BS 7671,

iii the equipment must disconnect itself from the distributor's network in the event of a network fault, and

iv the distributor must be advised of the installation before or at the time of commissioning.

Installations will need to meet the requirements of the Energy Networks Association Engineering Recommendation G83, *Recommendations for the connection of small-scale embedded generators (up to 16 A per phase) in parallel with the public low voltage distribution network.*

The requirements of the ESQC Regulations for small-scale embedded generators are discussed more fully in Chapter 15.

For generator installations larger than 16 A per phase see Energy Networks Association Engineering Recommendation G59-2.

10.4 Normative references

The following IEC standards are referred to in publication G83:

▶ IEC 61215: *Crystalline silicon terrestrial photovoltaic (PV) modules – Design qualification and type approval*

▶ IEC 61646: *Thin-film terrestrial photovoltaic (PV) modules - Design qualification and type approval*

▶ IEC 61730: *Photovoltaic (PV) module safety qualification*

▶ IEC 62446: *Grid connected photovoltaic systems - Minimum requirements for system documentation, commissioning tests and inspection*

Standards for equipment are being developed by IEC Committee TC 82.

10.5 Definitions

Part 2   (See also Figures 10.1 and 10.2.)

PV a.c. module: integrated module/convertor assembly where the electrical interface terminals are a.c. only. No access is provided to the d.c. side.

PV array: mechanically and electrically integrated assembly of PV modules, and other necessary components, to form a d.c. power supply unit.

PV array cable: output cable of a PV array.

PV array junction box: enclosure where PV strings of any PV array are electrically connected and where devices can be located.

PV cell: basic PV device which can generate electricity when exposed to light such as solar radiation.

PV convertor: device which converts d.c. voltage and d.c. current into a.c. voltage and a.c. current.

PV d.c. main cable: cable connecting the PV generator junction box to the d.c. terminals of the PV convertor.

PV generator: assembly of PV arrays.

PV generator junction box: enclosure where PV arrays are electrically connected and where devices can be located.

PV installation: erected equipment of a PV power supply system.

PV module: smallest completely environmentally protected assembly of interconnected PV cells.

PV string: circuit in which PV modules are connected in series, in order for a PV array to generate the required output voltage.

PV string cable: cable connecting PV modules to form a PV string.

PV supply cable: cable connecting the a.c. terminals of the PV convertor to a distribution circuit of the electrical installation.

Standard test conditions (STC): test conditions specified in BS EN 60904-3 for PV cells and PV modules.

Open-circuit voltage under standard test conditions U_{OC} STC: voltage under standard test conditions across an unloaded (open) generator or on the d.c. side of the convertor.

Short-circuit current under standard test conditions I_{SC} STC: short-circuit current of a PV module, PV string, PV array or PV generator under standard test conditions.

DC side (not a BS 7671 definition): part of a PV installation from a PV cell to the d.c. terminals of the PV convertor.

AC side (not a BS 7671 definition): part of a PV installation from the a.c. terminals of the PV convertor to the point of connection of the PV supply cable to the electrical installation.

Simple separation: separation between circuits or between a circuit and Earth by means of basic insulation.

10.6 Protection for safety

10.6.1 Protection against electric shock

712.410.3 PV equipment on the d.c. side is to be considered energized, even when the system is disconnected from the a.c. side.

Protection by extra-low voltage: SELV and PELV

712.414.1.1 For SELV and PELV systems, U_{OC} STC replaces U_0 and shall not exceed 120 V d.c.

10.6.2 Fault protection

712.411 **(a) Protection by automatic disconnection of supply**
Note: Protection by automatic disconnection of supply on the d.c. side requires special measures which are under consideration.

On the a.c. side, the PV supply cable shall be connected to the supply side of the protective device for automatic disconnection of circuits supplying current-using equipment.

712.411.3.2.1.2 Where an electrical installation includes a PV power supply system without at least simple separation between the a.c. side and the d.c. side, an RCD installed to provide fault protection by automatic disconnection of supply shall be type B to IEC 62423. Where the PV power supply convertor by construction is not able to feed d.c. fault currents into the electrical installation, an RCD of type B to IEC 62423 is not required. (A type B RCD ensures tripping for residual sinusoidal alternating currents, pulsating direct currents and smooth direct currents, whether suddenly applied or slowly rising. For special applications, refer to the manufacturer.)

The flow diagram below shows a typical decision-making process that could be used:-

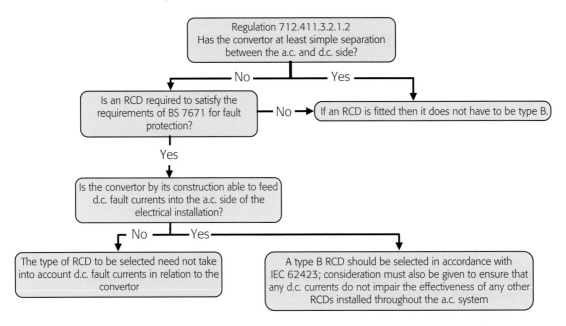

(b) Other protective measures

712.412 Protection by use of Class II or equivalent insulation should preferably be adopted on the d.c. side.

712.410.3.6 Protection by non-conducting location (Regulation 418.1) and earth-free local equipotential bonding (Regulation 418.2) are not permitted on the d.c. side.

10.6.3 Protection against overload on the d.c. side

Note: For PV cables on the d.c. side not complying with the paragraphs below, the requirements of Chapter 43 of BS 7671 apply for overload protection.

712.433.1 Overload protection may be omitted to PV string and PV array cables when the continuous current-carrying capacity of the cable is equal to or greater than 1.25 times I_{SC} STC at any location.

712.433.2 Overload protection may be omitted to PV main cables when the continuous current-carrying capacity of the PV main cable is equal to or greater than 1.25 times I_{SC} STC of the PV generator.

Note: These requirements are only relevant for protection of the cables. It is important to consult the manufacturer for the protection of the modules. To protect against the risk of module fires overload protection needs to be installed. The selection of overcurrent protective measures depends upon the system design and the number of strings. For advice on fuse selection refer to Edition 3 of the PV guide published by the Department for Business, Innovation and Skills (previously DTI guide). Also refer to IEC 60269-6 *Low-voltage fuses – Part 6: Supplementary requirements for fuse-links for the protection of solar photovoltaic energy systems.*

Refer to the manufacturer for more information.

10.6.4 Protection against fault current

712.434.1 The PV supply cable shall be protected against fault current by an overcurrent protective device installed at the connection to the a.c. mains.

10.6.5 Protection against electromagnetic interference (EMI)

712.444.4.4 To minimise voltages induced by lightning, the area of all wiring loops must be kept as small as possible.

10.7 Isolation and switching

712.537.2.1.1 To allow maintenance of the convertor, means of isolating the PV convertor from the d.c. side and the a.c. side shall be provided.

Note: Further requirements with regard to the isolation of a PV installation operating in parallel with the public supply system are given in Regulation 551.7.6 of BS 7671; see section 15.4 in Chapter 15.

10.8 Selection and erection of equipment

10.8.1 Compliance with standards

712.511.1 PV modules shall comply with the requirements of the relevant equipment standard, e.g. BS EN 61215 for crystalline PV modules. PV modules of Class II construction or with equivalent insulation are recommended if U_{OC} STC of the PV strings exceeds 120 V d.c.

The PV array junction box, PV generator junction box and switchgear assemblies shall be in compliance with BS EN 60439-1.

The designer and installer should verify that the cables are suitable for the installation conditions. Cable designed and tested specifically for use in PV systems is available.

An IEC standard for PV string cables is under development. It is recommended that in the interim cables should comply with UL 4703, or TUV 2 Pfg 1169 08.2007.

10.8.2 Operational conditions and external influences

712.512 Electrical equipment on the d.c. side shall be suitable for direct voltage and direct current.

PV modules may be connected in series up to the maximum allowed operating voltage of the PV modules or the PV convertor, whichever is lower. Specifications for this equipment shall be obtained from the equipment manufacturer.

If blocking diodes are used, their reverse voltage shall be rated for 2 x U_{OC} STC of the PV string. The blocking diodes shall be connected in series with the PV strings (see Figures 10.1 and 10.2).

712.512.2.1 When installing PV modules the installer must follow the manufacturer's instructions for mounting so that adequate heat dissipation is provided under the conditions of maximum solar radiation to be expected. Such instructions are required by the equipment standard.

10.8.3 Accessibility

712.513.1 The selection and erection of equipment shall facilitate safe maintenance and shall not adversely affect provisions made by the manufacturer of the PV equipment to enable maintenance or service work to be carried out safely.

10.8.4 Wiring systems

712.522 **Selection and erection in relation to external influences**

PV string cables, PV array cables and PV d.c. main cables shall be selected and erected so as to minimise the risk of earth faults and short-circuits.

Note: This may be achieved for example by reinforcing the protection of the wiring against external influences by the use of single-core sheathed cables.

Wiring systems shall withstand the expected external influences such as wind, ice formation, temperature and solar radiation.

10.8.5 Devices for isolation and switching

712.537.2.2.1 In the selection and erection of devices for isolation and switching to be installed between the PV installation and the public supply, the public supply shall be considered the source and the PV installation shall be considered the load. A notice warning of two

514.15.1 sources of supply should be provided at each point of isolation.

712.537.2.2.5 A switch disconnector shall be provided on the d.c. side of the PV convertor.

All junction boxes (PV generator and PV array boxes) shall carry a warning label indicating that active parts inside the boxes may still be live after isolation from the PV convertor.

WARNING
PV SYSTEM
Parts inside this box or enclosure may still be live after isolation from the supply.

10.9 Earthing arrangements and protective conductors

712.54   Where protective bonding conductors are installed, they shall be parallel to and in as close contact as possible with d.c. cables and a.c. cables and accessories.

The flow diagram below shows a typical decision-making process that could be used:-

10.9.1 Types of system earthing

712.312.2   Earthing of one of the live conductors of the d.c. side is permitted, if there is at least simple separation between the a.c. side and the d.c. side.

Note: Any connections with Earth on the d.c. side should be electrically connected so as to avoid corrosion.

Engineering Recommendation G83 published by the Energy Networks Association allows the connection of one pole of the d.c. side of the convertor to the DNO's earth terminal subject to certain conditions. Refer to the manufacturer for more information.

A risk assessment should be carried out to determine exposed-conductive-parts and extraneous-conductive-parts.

▶ **Figure 10.1** PV installation – general schematic, one array

▶ **Figure 10.2** PV installation – example with two or more arrays

Exhibitions, shows and stands   11

11.1 Introduction

Introduction to Amendment No 1 to the 17th Edition

711.411.4 Under 'Protection against electric shock', TN system. Regulation 711.411.4 has been modified and now refers to PME.

Harmonized standards are now referenced and changes have been made to the standards under wiring systems.

Introduction to the 17th Edition

The 16th Edition did not include requirements for exhibitions, shows and stands, but guidance was given in the previous edition of this Guidance Note. However, new Sect 711 Section 711 concerning the electrical installations of exhibitions, etc. has been introduced into the 17th Edition.

Section 711 specifies particular requirements for this type of electrical installation relating to assessment of general characteristics, protection against electric shock, protection against thermal effects, selection and erection of equipment and inspection and testing.

11.2 Scope

Part 2 This chapter is concerned with temporary electrical installations in exhibitions, shows and stands including mobile and portable displays and equipment. Such installations may be installed indoors or outdoors within permanent or temporary structures. It does not apply to the fixed electrical installation of the building, if any, in which the exhibition, show or stand takes place.

Section 711 of BS 7671:2008 does not apply to electrical systems as defined in BS 7909 used in structures, sets, mobile units etc. as used for public or private events, touring shows, theatrical, radio, TV or film productions and similar activities of the entertainment industry.

11.3 The risks

The particular risks associated with exhibitions, shows and stands are those of electric shock and fire. These arise from:

1 the temporary nature of the installation
2 lack of permanent structures
3 severe mechanical stresses
4 access to the general public.

Because of these particular risks, additional measures are recommended as described in the following sections.

11.4 Protection against electric shock

711.410.3.4 A cable intended to supply temporary structures shall be protected at its origin by an RCD whose rated residual operating current does not exceed 300 mA. This device shall provide a delay by using a device in accordance with BS EN 60947-2, or be of the S-type in accordance with BS EN 61008-1 or BS EN 61009-1 for discrimination with RCDs protecting final circuits. See Figure 11.1.

> **Note:** The requirement for additional protection relates to the increased risk of damage to cables in temporary locations.

711.410.3.5 Protective measures against electric shock by means of obstacles and by placing out of reach (Section 417) are not permitted.

711.410.3.6 The protective measures of non-conducting location (Regulation 418.1) and earth-free local equipotential bonding (Regulation 418.2) are not permitted.

11.4.1 Protection by automatic disconnection of supply

711.411.4 **TN system**

Except for a part of an installation within a building, a PME earthing facility shall not be used as the means of earthing for an installation falling within the scope of Section 711 except:

i where the installation is continuously under the supervision of a skilled or instructed person, and

ii the suitability and effectiveness of the means of earthing has been confirmed before the connection is made.

Where these conditions cannot be met, if the exhibition supply is derived from a building that is supplied by a PME system then the exhibition supply will have to be part of a TT system having a separate connection to Earth independent from the PME earthing.

11.4.2 Additional protection

711.411.3.3 Each socket-outlet circuit not exceeding 32 A and all final circuits other than for emergency lighting shall be protected by an RCD having the characteristics specified in Regulation 415.1.1. However, consideration has to be given to the hazards of loss of lighting in such a public place, particularly when crowded. Lighting of such areas should always be on at least two separate circuits with separate RCDs, and should preferably be out of reach of the general public (see Figure 11.1).

▼ Figure 11.1 Exhibition/show distribution with standby generator

Note: Regulation 21 of the ESQC Regulations has requirements for switched alternative sources of energy; see section 11.9.

11.4.3 Bonding of stands, vehicles, wagons, caravans and containers

711.411.3.1.2 Structural metallic parts which are accessible from within the stand, vehicle, wagon, caravan or container shall be connected through the main protective bonding conductors to the main earthing terminal within the unit.

11.4.4 Protection by SELV or PELV

711.414.4.5 Where SELV or PELV installations are used, whatever the nominal voltage, basic protection is required to be provided by basic insulation or by barriers or enclosures, providing protection of at least IPXXD or IP4X.

11.5 Protection against thermal effects

Chap 42 There is often an increased risk of fire and burns in temporary locations. For this reason it is important to comply with all the relevant requirements of Chapter 42.

11.5.1 Protection against fire

Installation designers must bear in mind that stored materials may present a particular hazard in such a location, particularly fodder, cardboard boxes etc.

422.3.7
552.1.2 Motors which are automatically or remotely controlled and are not continuously supervised should be fitted with manual reset devices for protection against excess temperature, accessible only to skilled persons.

11.5.2 Lighting

711.422.4.2
422.3.1
559.5.1 Lighting equipment such as incandescent lamps, spotlights and small projectors, and other equipment or appliances with high temperature surfaces should, in addition to being suitably guarded, be arranged well away from combustible materials such as

to prevent contact. Equipment should be installed and located in accordance with relevant standards and manufacturers' instructions.

711.559.5 Luminaires that are mounted below a height of 2.5 m from floor or ground level or mounted in a position that is accessible to accidental contact, should be firmly and adequately fixed, and sited or guarded so as to prevent the risk of injury to persons or ignition of materials. Where this includes outdoor lighting installations the requirements 559.10 of Regulation group 559.10 would apply.

Showcases and signs should be constructed from materials having adequate heat resistance, mechanical strength, electrical insulation and ventilation, taking account of the combustibility of exhibits in relation to the heat generated. The manufacturer's instructions must be complied with.

Where there is a concentration of electrical equipment, including luminaires, that might generate considerable heat, adequate ventilation must be provided.

GN4 Similar guidance for protection against fire for this type of location can be found in section 8.3 of Guidance Note 4: *Protection Against Fire*.

11.6 Isolation

711.537.2.3 Every separate temporary structure, such as vehicles, stands or units, intended to 537.2 be occupied by one specific user, and each distribution circuit supplying outdoor installations, should be provided with its own readily accessible and properly identifiable means of isolation. Switches, circuit-breakers and residual current devices etc. considered suitable for isolation by the relevant standard or the manufacturer may be used.

11.7 Measures of protection against overcurrent

430.3 All circuits should be protected against overcurrent by a suitable protective device located at the origin of the circuit.

11.8 Selection and erection of equipment

711.51 Control and protective switchgear should be placed in closed cabinets which can only be opened by a key or tool, except those parts which are designed and intended to Part 2 be operated by ordinary persons.

512.2 Equipment, particularly switchgear and fusegear, must be mounted away from locations that may not be weatherproof. Tent poles etc. where used for mounting switchgear are often the weak point in the weather tightness of temporary structures.

The means of isolation for each stand or unit should not be locked away and should be readily accessible and obvious to the stand user (see section 11.6).

11.8.1 Wiring systems

Sect 522 Particular care must be paid to the selection and installation of cables to ensure that Sect 523 the mechanical protection, insulation, heat resistance and current-carrying capacity are sufficient for the conditions likely to be encountered.

711.52 Wiring cables shall be copper, have a minimum cross-sectional area of 1.5 mm^2, and shall comply with an appropriate British or Harmonized Standard for either thermoplastic or thermosetting insulated electric cables. Flexible cables shall not be laid in areas accessible to the public unless they are protected against mechanical damage.

Mechanical protection or armoured cables should be used wherever there is a risk of mechanical damage.

Underground cables are susceptible to damage by structure support pins which may be up to 1 m in length. Exhibitors must be advised of the presence of cables and, if necessary, the cable route marked. The general rules for buried cables must be followed; see also section 13.3 in Chapter 13.

711.521 Where a building is used for exhibitions etc. and does not include a fire alarm system, either of the following cable systems should be used:

▶ flame retardant cables to BS EN 60332-1-2 or BS EN 60332-3, and low smoke to BS EN 61034-2
▶ single-core or multicore cables enclosed in metallic or non-metallic conduit or trunking that provides fire protection in accordance with BS EN 61386 series or BS EN 50085 series. This is to provide a degree of protection of at least IP4X.

11.8.2 Electrical connections

711.526.1 Joints should not be made in cables, except as a connection into a circuit.

Conductor connections should be within an enclosure providing a degree of protection of at least IPXXD or IP4X and the enclosure should incorporate a suitable cable anchorage.

11.8.3 Lighting installations

ELV lighting systems for filament lamps

711.559.4.2 Extra-low voltage systems for filament lamps must comply with BS EN 60598-2-23.

Lampholders

711.559.4.3 Insulation-piercing lampholders should not be used unless the cables and lampholders are compatible and the lampholders are non-removable once fitted to the cable.

Electric discharge lamp installations

711.559.4.4 Installations of any luminous tube, sign or lamp as an illuminated unit on a stand, or as an exhibit, with nominal power supply voltage higher than 230/400 V a.c., must comply with Regulations 711.559.4.4.1 to 711.559.4.4.3.

The sign or lamp should be installed out of arm's reach or be adequately protected to reduce the risk of injury to persons.

The facia or stand fitting material behind luminous tubes, signs or lamps must be non-ignitable.

Emergency switching device

711.559.4.4.3 A separate circuit should be used to supply signs, lamps or exhibits, which is controlled by an emergency switch. The switch must be easily visible, accessible and clearly marked.

Protection against thermal effects

711.559.5 Luminaires mounted below 2.5 m (arm's reach) from floor level or otherwise accessible to accidental contact must be firmly and adequately fixed, and so sited or guarded as to prevent risk of injury to persons or ignition of materials.

> **Note:** In the case of outdoor lighting installations, Section 559 also applies, and a degree of protection of at least IP33 may be required.

11.8.4 Electric motors

711.55.4.1
537.4.2.1 Where an electric motor might give rise to a hazard the motor must be provided with an effective means of isolation on all poles and where necessary an emergency stop, and the means should be adjacent to the motor it controls (see BS EN 60204-1).

11.8.5 Extra-low voltage transformers and electronic convertors

711.55.6 A manual reset protective device should protect the secondary circuit of each transformer or electronic convertor.

Particular care should be taken to install ELV transformers out of reach of the public and adequate ventilation should be allowed for. Access by a competent person for testing and by a skilled person competent in such work for maintenance shall be provided.

Electronic convertors should conform with BS EN 61347-1.

11.8.6 Socket-outlets

711.55.7 An adequate number of socket-outlets should be installed to allow user requirements to be met safely.

Floor mounted sockets should preferably not be installed, but where their use is unavoidable they should be adequately protected from mechanical damage and ingress of water.

11.9 Generators

Sect 551 Installations incorporating generator sets must comply with Section 551 of BS 7671. Where a generator is used to supply the temporary installation using a TN or TT system, it must be ensured that the installation is earthed, preferably by separate earth electrodes. For TN systems all exposed-conductive-parts should be bonded back to the generator. The neutral conductor and/or star point of the generator should be connected to the exposed-conductive-parts of the generator and reference earthed.

Part VI of the ESQC Regulations provides requirements for generation. Regulation 21 has requirements for switched alternative sources of energy (see Figure 11.1) as follows:

Where a person operates a source of energy as a switched alternative to a distributor's network, he shall ensure that that source of energy cannot operate in parallel with that network and where the source of energy is part of a low voltage consumer's installation, that installation shall comply with British Standard Requirements [meaning BS 7671].

The requirements for parallel operation are much more onerous.

11.10 Safety services

Chapter 56
Where an exhibition is held within a building, it is assumed that the emergency lighting and/or fire safety systems etc. will be part of the permanent installation within that building. Care should be taken to ensure that existing emergency escape signs and escape routes are not obscured, impeded or blocked.

Additional emergency lighting should be installed in those areas not covered by the permanent installation. Where an exhibition is constructed out-of-doors, an adequate fire alarm system should be installed to enclosed areas to facilitate emergency evacuation.

Where an event is taking place out-of-doors and is open to the public in partial or total darkness, then:

▶ emergency lighting should be provided to escape routes
▶ provision should be made to ensure that alternative sources of supply for general lighting sufficient for safe evacuation are available throughout the area.

Reference should be made to the following British Standards:

▶ BS 5266 *Emergency lighting*
▶ BS 5839 *Fire detection and fire alarm systems for buildings.*

11.11 Inspection and testing

711.6
The temporary electrical installations should be retested on site in accordance with Chapter 61 of BS 7671, after each assembly on site.

Users (e.g. exhibitors and stall holders) should be advised to visually check electrical equipment for damage on a daily basis.

Floor and ceiling heating systems

12

[Illustration courtesy of Warmup plc]

12.1 Introduction

Amendment No 1 to the 17th Edition

Sect 753 There are no significant changes in Section 753 as a result of Amendment No 1.

The changes introduced by the 17th Edition (BS 7671:2008) when it was published in January 2008 are described below for information.

Introduction to the 17th Edition

This is a new section that has been introduced into the 17th Edition.

753.411.3.2 Regulation 753.411.3.2 has been included so that 30 mA RCDs are required for ceiling or floor heating systems to protect against penetration of the heating elements.

There are requirements to protect the building structure against fire and to avoid the overheating of the heating system.

Heating units manufactured without exposed-conductive-parts must be provided on site with a metal grid.

12.2 Scope

753.1 Section 753 applies to the installation of electric floor and ceiling heating systems which are erected as either thermal systems (off-peak) or direct heating systems. It does not apply to the installation of wall heating systems. Heating systems for use outdoors are not considered.

Note 1: A ceiling located under the roof of a building down to a vertical height of 1.50 m measured from the finished floor surface is also regarded as a ceiling within the meaning of the Regulations.

Note 2: Currently wall heating systems are not included in Section 753 of BS 7671. Persons involved in this work should seek manufacturers' advice.

12.3 The risks

753.415.1 The risks associated with ceiling heating systems are penetration of the heating elements by nails, drawing pins, etc. pushed through the ceiling surface. For this reason, additional protection is required by the use of a 30 mA RCD.

Similarly, there are concerns that underfloor heating installations can be damaged by carpet gripper nails, etc. and for similar reasons protection by a 30 mA RCD is required.

There is a risk of fire to the building structure due to the use of heating, therefore there are requirements to avoid overheating of the floor or ceiling heating system.

12.4 Protection against electric shock

12.4.1 Locations containing a bath or shower

701.753 Underfloor heating installations in locations containing a bath or shower should have an overall earthed metallic grid or the heating cable should have an earthed metallic sheath, which must be connected to the protective conductor of the supply circuit; see also 12.10.

12.4.2 Other areas

753.411.3.2 In areas other than special locations, Class I heating units which do not have an exposed-conductive-part, i.e. integrated earth screen or sheath, must have a metallic grid, with a spacing of not more than 30 mm, installed above the floor heating elements. The grid must be connected to the protective conductor of the electrical installation and the heating system protected by an RCD with a rated residual operating current not exceeding 30 mA.

753.415.1 In areas where occupants are not expected to be completely wet, a circuit supplying heating equipment of Class II construction or equivalent insulation must be provided with additional protection by the use of an RCD with a rated residual operating current not exceeding 30 mA.

Note: Limitation of the rated heating power to 7.5 kW/230 V or 13 kW/400 V downstream of a 30 mA RCD may avoid unwanted tripping due to leakage capacitance. Values of leakage capacitance may be obtained from the manufacturer of the heating system.

12.5 Protection against overheating

12.5.1 Protection against burns

753.423

In floor areas where contact with skin or footwear is possible, the surface temperature of the floor shall be limited (for example, 35 °C).

12.5.2 Protection against overheating

753.424.3.1

To avoid overheating of floor or ceiling heating systems in buildings, at least one of the following measures shall be applied to limit the temperature and the heating zone to a maximum of 80 °C:

i   Appropriate design of the heating system
ii   Appropriate installation of the heating system
iii Use of protective devices.

Heating units shall be connected to the electrical installation via cold tails or suitable terminal fittings.

Heating units shall be inseparably connected to the cold tails, e.g. by welding, brazing or by compression jointing techniques.

Heating units must not cross expansion joints.

GN4

Additional guidance for protection against fire for this type of location can be found in section 8.5 of Guidance Note 4: *Protection Against Fire*.

753.424.3.2

As the heating unit may cause higher temperatures or arcs under fault conditions, special measures to meet the requirements of Chapter 42 should be taken when the heating unit is installed close to easily ignitable building structures, such as placing on a metal sheet, in metal conduit or at a distance of at least 10 mm in air from the ignitable structure.

12.6 Standards

753.511

Flexible sheet heating elements should comply with BS EN 60335-2-96 and heating cables should comply with IEC 60800.

12.7 External influences

753.512.2.5

Heating units for installation in ceilings shall be at least IPX1 and heating units for installation in floors of concrete or similar material shall be not less than IPX7 with appropriate mechanical properties.

12.7.1 Operational conditions

753.512.1.6

Precautions shall be taken not to stress the heating unit mechanically; for example, the material by which it is to be protected in the finished installation shall cover the heating unit as soon as possible.

12.8 Identification

753.514   The designer of the installation/heating system or installer shall provide a plan for each heating system, containing the following details:

1 Manufacturer and type of heating units
2 Number of heating units installed
3 Length/area of heating units
4 Rated power
5 Surface power density
6 Layout of the heating units in the form of a sketch, a drawing or a picture
7 Position/depth of heating units
8 Position of junction boxes
9 Conductors, shields and the like
10 Heated area
11 Rated voltage
12 Rated resistance (cold) of the heating units
13 Rated current of overcurrent protective devices
14 Rated residual operating current of RCD
15 The insulation resistance of the heating installation and the test voltage used
16 The leakage current.

The plan shall be fixed to, or adjacent to, the distribution board of the heating system.

Section 12.11 describes advice to be provided by the installer for the user of the installation.

12.9 Heating-free areas

753.522.4.3   Where heating units are installed, there shall be heating-free areas where drilling and fixing by screws, nails and the like are permitted. The installer shall inform other contractors that no penetrating means, such as screws for door stoppers, shall be used in the area where floor or ceiling heating units are installed.

753.520.4   It may be necessary to provide areas of floor or ceiling that are unheated, e.g. where fixtures to the floor or ceiling would prevent the proper emission of heat.

753.522.1.3   Account shall be taken of the increase in ambient temperature and of its effect upon the cables, including cold tails (circuit wiring) and control wiring installed in heated   zones.

12.10 Locations containing a bath or shower and swimming pools and other basins

701.753
702.55.1   Where underfloor heating systems are installed in bathrooms and swimming pools supplied at voltages other than SELV, then the heating element should be provided either with a metallic sheath or overall screen or a metallic grid positioned above the heating elements. The screen or grid shall be connected to the supplementary bonding and earthing for the facility. In addition, the supply to the heating elements should be controlled via an RCD with a rated residual operating current not exceeding 30 mA with no adjustable time delay.

As such facilities are likely to suffer from the ingress of water via tiling etc. and also from corrosion, provision for ready access to all terminations to permit the thorough testing of such installations is vital. The use of suitable plastic materials for enclosures and conduit/trunking etc. is one method of minimising potential corrosion effects.

Furthermore, regular visual inspection of these installations, particularly in swimming pools where corrosion from the chemicals is more likely, should be carried out and adequate records kept.

Sects 701/702 All such facilities must comply with, as applicable, Sections 701 and/or 702 of BS 7671:2008.

12.11 Information from the contractor for the user of the installation

Fig 753 A description of the heating system shall be provided by the installer/contractor of the heating system to the owner of the building or his/her agent upon completion of the installation.

The **description** shall contain at least the following information:

a Description of the construction of the heating system, which must include the installation depth of the heating units;

b Location diagram with information concerning
▶ the distribution of the heating circuits and their rated power;
▶ the position of the heating units in each room;
▶ conditions which have been taken into account when installing the heating units, e.g. heating-free areas, complementary heating zones, unheated areas for fixing means penetrating into the floor covering;

c Data on the control equipment used, with relevant circuit diagrams and the dimensioned position of floor temperature and weather conditions sensors, if any;

d Data on the type of heating units and their maximum operating temperature.

The installer/contractor shall inform the owner that the description of the heating system includes all necessary information, e.g. for repair work.

The designer/installer/contractor of the heating system shall hand over an appropriate number of instructions for use to the owner or his/her agent upon completion. One copy of the instructions for use shall be permanently fixed in or near each relevant distribution board.

The **instructions for use** shall include at least the following information:

a Description of the heating system and its function;

b Operation of the heating installation in the first heating period in the case of a new building, e.g. regarding drying out;

c Operation of the control equipment for the heating system in the dwelling area and the complementary heating zones as well, if any;

d Information on restrictions on placing of furniture or similar. Information provided to the owner shall cover the restrictions, if any, including:
▶ whether additional floor coverings are permitted, for example, carpets with a thickness of >10 mm may lead to higher floor temperatures which can adversely affect the performance of the heating system
▶ where pieces of furniture solidly covering the floor and/or built-in cupboards may be placed on heating-free areas
▶ where furniture, such as carpets, seating and rest furniture with pelmets, which in part do not solidly cover the floor, may not be placed in complementary heating zones, if any;

e Information on restrictions on placing of furniture or similar;

f   In the case of ceiling heating systems, restrictions regarding the height of furniture. Cupboards of room height may be placed only below the area of ceiling where no heating elements are installed;

g   Dimensioned position of complementary heating zones and placing areas;

h   Statement that, in the case of thermal floor and ceiling heating systems, no fixing shall be made into the floor and ceiling respectively. Excluded from this requirement are unheated areas. Alternatives shall be given, where applicable.

Gardens (other than horticultural installations)

13

13.1 Introduction

Persons carrying out electrical work in dwellings must comply with the Building Regulations of England and Wales, in particular Part P (Electrical safety – dwellings).

Part P of the Building Regulations do not apply in Scotland or Northern Ireland. In Scotland the requirements of the Building Regulations (Scotland) 2004 apply, in particular Regulation 9, and in Northern Ireland the Building Regulations (Northern Ireland) 2000 (as amended) apply.

Introduction to the 17th Edition incorporating Amendment No 1:2011

Gardens and similar installations are not included as special locations in BS 7671. However, electrical installation work in these locations is notifiable under Part P (Electrical safety – dwellings) of the Building Regulations 2000 in England and Wales. Specifically, this applies in a garden or in or on land associated with a building where the electricity is from a source located within or shared with a dwelling.

The general requirements of BS 7671:2008(2011) for outdoor circuits and equipment apply to all gardens.

13.2 The risks

Gardens can present a number of potential risks where electrical installations are concerned. These risks may include:

▶ contact of persons with the general mass of Earth, possibly with bare feet
▶ a frequently wet environment
▶ the wearing of minimal clothing
▶ gardening activity that may cause damage or disturbance to cables or equipment
▶ insertion of spikes in the ground for securing marquees, inflatables etc.

In public gardens it should be recognised that electrical equipment, e.g. lighting, may be accessible to the general public.

13.3 Cables

13.3.1 Buried cables

522.8.10 Cables should be protected against foreseeable damage, either by armouring or by suitable enclosure. Unprotected cables should not be buried direct in the ground nor should they be clipped to wooden fences, etc. which may provide inadequate support and protection.

Problems arise when either ground levels are lowered so that cables have insufficient cover or when ground levels are raised so that cables which were not intended to be buried and are not suitable for burial become buried. Such problems can arise during the course of a project and the intended ground level should be formally ascertained before the cables are installed. It must be remembered that the layout of a garden can be changed totally within a few seasons and great care must be taken to route cables where they are not likely to be disturbed or damaged, e.g. around the edge of the plot and at sufficient depth.

Buried cable routes should be identified by local route markers, and recorded on drawings; cables should be buried at least 450 mm (preferably deeper) below the lowest local ground level, and a route marker tape laid along the cable route about 150 mm below the surface.

Note: IEC 61386-24 is the standard for underground conduits.

13.3.2 Cables exposed to sunlight

522.2.1
522.11.1

Cables should be shielded from prolonged exposure to direct sunlight, or be of a type suitable for such exposure. Cable with a black sheath is recommended. Generally, the ultraviolet light from the sun will affect plastics and the cable manufacturer's advice should be taken. However, cables must not be so enclosed that heat dissipation is inadequate.

13.3.3 Cables taken overhead

GN1

Cables taken overhead should have a suitable rigid support or a catenary wire, or be a type suitable for such installation. See Guidance Note 1, Appendix G.3 for more information.

13.4 Socket-outlets

411.3.3

Mobile equipment with a current rating not exceeding 32 A for use outdoors requires additional protection by means of a 30 mA RCD with characteristics in accordance with Regulation 415.1.1. Socket-outlets that can be expected to be used to supply this equipment outdoors will therefore require this additional protection, which can be provided by using an RCCB complying with BS 61008-1 or BS 4293, an RCBO complying with BS EN 61009-1 or an SRCD complying with BS 7288.

Socket-outlets should be suitably placed to be convenient for their purpose, and of IP rating e.g. IP44 or better if located outdoors.

411.3.3 note 2

An RCD is not required for a socket-outlet in this instance if the circuit is reduced low voltage or extra-low voltage.

13.5 Fixed equipment

Fixed equipment in the garden, such as permanently fixed garden lighting, should be securely fixed with all cables buried, or supported away from the ground or paths. All-insulated Class II equipment is recommended where possible for increased safety. Decorative lighting, including 'festoon' type lighting, should be permanently fixed if used

Sect 559

regularly. See Section 559 of BS 7671:2008 incorporating Amendment No 1:2011.

Other fixed equipment, including pumps, etc., should be installed in accordance with the manufacturer's instructions and the general requirements of BS 7671:2008.

13.6 Ponds

Sect 702 In view of the risk of accidental or intentional immersion it is recommended that the same rules should be applied to garden ponds, especially larger ones, as are applied to swimming pools. Equipment (including cables) must be suitable for the purpose and of a suitable IP rating, or be installed in a suitable enclosure. Class II equipment should be utilised where possible. Cables should be installed in ducts or conduits built into the pond structure and not allowed to lie loose around the area. All connections must be made in robust, watertight junction boxes. Equipment to IP55 or better is recommended.

Pond lighting should meet the requirements of BS EN 60598-2-18, pumps BS EN 60335-2-41, and other equipment BS EN 60335-2-55.

13.7 Supplies to outbuildings within the garden

The following requirements apply when supplying a detached garage or shed from a dwelling. The preferred method is to supply the detached outbuilding from a spare way in the consumer unit in the dwelling and to install a small distribution board in the outbuilding (see Figure 13.1). A suitably sized circuit-breaker should be fitted in a spare way in the dwelling's consumer unit and a submain cable installed to supply the consumer unit in the garage. The cable run from the house to the garage must be suitably protected, see section 13.3. The consumer unit in the garage should be fitted with circuit-breakers for the final circuits, such as 6A for the lighting circuit and 16 A for the socket-outlet circuit.

13.7.1 Additions and alterations to an existing installation

132.16 Where work is being carried out to an existing installation Regulation 132.16 requires that the rating and condition of existing equipment, including that of the distributor, should be adequate for the additional load and that the existing earthing and bonding arrangements are also adequate.

13.7.2 Earthing and main equipotential bonding

411.3.1.2
54 Where an installation serves a detached building which contains extraneous-conductive-parts, Regulation 411.3.1.2 requires main protective bonding conductors complying with Chapter 54 to connect to the main earthing terminal (in the house) extraneous-conductive-parts including the following:

i   Water installation pipes

ii   Gas installation pipes

iii   Other installation pipework and ducting

iv   Central heating and air conditioning systems

v   Exposed metallic structural parts of the building.

Table 54.8 For an outbuilding fed from a dwelling having a PME supply and where the outbuilding includes extraneous-conductive-parts, a main protective bonding conductor in accordance with Table 54.8 will be required (at least 10 mm^2 copper) to be run between the dwelling's main earthing terminal and the extraneous-conductive-parts in the outbuilding.

Note: If there are no extraneous-conductive-parts in the outbuilding then main equipotential bonding will not be required.

13.7.3 TT system

An alternative possibility is to make the small installation in the garage part of a TT system. A local earth electrode must be provided at the garage. To achieve protection by automatic disconnection of supply for both circuits in the garage an RCD must be employed. Main equipotential bonding will need to be provided at the garage connecting the metal water pipe and any other extraneous-conductive-parts to the earthing terminal in the small distribution board. An exposed-conductive-part connected to one means of earthing must not be simultaneously accessible with an exposed-

411.3.1.1 conductive-part connected to another means of earthing (Regulation 411.3.1.1 refers). Where the installation in the garage is supplied by an armoured cable, the armour or any protective conductor in the cable must not be connected to and must not be simultaneously-accessible with any exposed-conductive-parts in the outbuilding.

13.7.4 RCD protection

411.3.3 Regulation 411.3.3 requires additional protection by means of a 30 mA RCD to be provided for socket-outlets not exceeding 20A for use by ordinary persons. Certain exceptions are permitted. The regulation also requires that additional protection by an RCD to be provided in circuits likely to supply mobile equipment for use outdoors.

RCD protection can be provided as follows:

▶ An RCBO in the dwelling's consumer unit for the garage supply, or
▶ An RCCB in the dwelling's consumer unit, or
▶ Selecting a device that includes RCD protection as the main switch for the small two-way consumer unit in the garage, or
▶ Protecting the socket-outlet circuit in the garage with an RCBO, or
▶ Providing an SRCD (a socket-outlet incorporating RCD protection) for the socket-outlet in the garage.

13.7.5 Inspection, testing and certification

Inspection and testing must be performed to confirm the adequacy of the relevant parts of the existing installation which will support the changed requirements, the upgrading of the existing installation necessary to support the addition or alteration, and the addition or alteration itself. The requirements for initial verification are contained

Part 6 in Part 6 of BS 7671 and further information on the requirements for inspection and
GN 3 testing is given in Guidance Note 3 *Inspection & Testing*. Compliance with BS 7671 must be verified for every addition or alteration. Requirements for certification and reporting in respect of electrical installations are also given in Part 6. An Electrical Installation Certificate must be provided to the person ordering the work giving details of the extent of the installation covered by the certificate, together with a record of the inspection and the results of the testing.

▼ Figure 13.1
Outbuilding supply
taken from a spare way
in the consumer unit

13.8 Supplies to hot tubs

Persons using hot tubs are particularly at risk due to:

1 lack of clothing, particularly footwear

2 presence of water reducing contact resistance

3 immersion in water, reducing total body resistance

4 ready availability of earthed metalwork

5 increased contact area

6 contact with ground where outdoors.

There are no specific requirements in BS 7671 for electrical installations associated with hot tubs. However, supplies to hot tubs should be protected by a 30 mA RCD.

701 Where a hot tub is located in a room other than a bathroom, for example a garden room/shed/etc, it is recommended that the requirements of Section 701 should be applied in full. The requirements are given in Chapter 1 of this guide.

702 Where a hot tub is located outdoors in the open air it is recommended that the requirements of Section 702 should be applied in full. The requirements are given in Chapter 2 of this guide.

Persons involved in the design of the electrical installation for a hot tub should consult the product standard BS EN 60335-2-60: *Specification for safety of household and similar appliances: Particular requirements for whirlpool baths and whirlpool spas.*

Mobile or transportable units

14

14.1 Introduction

Introduction to Amendment No 1 to the 17th Edition

717 Section 717 of Amendment No 1 to BS 7671:2008 includes a number of notable changes.

The scope has been widened to include both a.c. and d.c. and reference is made to BS 7909.

717.411.4 Requirements covering supplies have been added. Regulation 717.411.4 has been modified, and now refers to PME. The requirements covering identification, wiring systems, and other equipment have been expanded.

Introduction to the 17th Edition

The 16th Edition did not include requirements for mobile or transportable units, but guidance was given in the previous edition of this Guidance Note. However, new Section 717 concerning the electrical installations of mobile or transportable units has been introduced into the 17th Edition.

Section 717 specifies particular requirements for this type of electrical installation relating to protection against electric shock and selection and erection of equipment.

14.2 Scope

717 The particular requirements of Section 717 are applicable to mobile or transportable units. For the purposes of this section, the term 'unit' is intended to mean a vehicle and/or mobile or transportable structure in which all or part of an electrical installation is contained, which is provided with a temporary supply by means of, for example, a plug and socket-outlet.

Units are either:

   i of the mobile type, e.g. vehicles (self-propelled or towed), or
   ii of the transportable type, e.g. containers or cabins.

Examples of the units include technical and facilities vehicles for the entertainment industry, medical or health screening services, advertising, firefighting, workshops, offices, transportable catering units etc. The requirements are not applicable to:

a   generating sets (Section 551 of BS 7671)
b   marinas and pleasure craft (Section 709)
c   mobile machinery in accordance with BS EN 60204-1
d   caravans (Section 721)
e   traction equipment of electric vehicles
f   electrical equipment required by a vehicle to allow it to be driven safely or used on the highway.

The guidance given here is in addition to the general requirements of BS 7671. Where other special locations such as rooms containing showers or medical locations form part of a mobile or transportable unit, the special requirements for those installations should also be taken into consideration. In this regard, particular reference should be made to Part 7 of BS 7671 and the other relevant chapters of this Guidance Note.

14.3 The risks

The risks associated with mobile and transportable units arise from:

i   risk of loss of connection to earth, due to use of temporary cable connections and long supply cable runs; the repeated use of cable connectors which may give rise to 'wear and tear' and the potential for mechanical damage to these parts
ii   risks arising from the connection to different national and local electricity distribution networks, where unfamiliar supply characteristics and earthing arrangements are found
iii   impracticality of establishing an equipotential zone external to the unit
iv   open-circuit faults of the PEN conductor of PME supplies raising the potential of all metalwork (including that of the unit) to dangerous levels
v   risk of shock arising from high functional currents flowing in protective conductors – usually where the unit contains substantial amounts of electronics or communications equipment
vi   vibration while the vehicle or trailer is in motion, or while a transportable unit is being moved – causing faults within the unit installation.

Particular requirements to reduce these risks include:

a   checking the suitability of the electricity supply before connecting the unit
b   accessible conductive parts of the unit to be connected through the main equipotential bonding to the main earthing terminal within the unit
c   a regime of regular inspection and testing of connecting cables and their couplers, supported by a logbook system of record keeping
d   the use of flexible cables
e   the use of RCDs
f   clear and unambiguous labelling of units, indicating types of supply which may be connected
g   particular attention paid to the maintenance and periodic inspection of installations.

14.4 Supplies

717.313   ### 14.4.1 General

One or more of the following methods must be used to supply a unit:

Fig 717.1   i   Connection to a low voltage generating set in accordance with Section 551 (see
Fig 717.2   Figures 14.1 and 14.2)

Fig 717.3

ii Connection to a fixed electrical installation in which the protective measures are effective (see Figure 14.3)

Fig 717.4
Fig 717.6
Fig 717.7

iii Connection through means providing simple separation, in accordance with Section 413, from a fixed electrical installation (see Figures 14.4 to 14.6)

717.411.4

NOTE 1: In cases i, ii and iii, an earth electrode may be provided where supplies are used external to the vehicle (see Regulation 717.411.4).

717.411.6.2(ii)

NOTE 2: In the case of Figure 14.4, an earth electrode may be necessary for protective purposes (see Regulation 717.411.6.2(ii)).

NOTE 3: Simple separation or electrical separation is appropriate, for example, where information technology equipment is used in the unit or where a reduction of electromagnetic disturbances is necessary, or if high leakage currents are to be expected (use of frequency convertors), and/or if the supply to the unit comes from alternative supply systems (as is the case in disaster management).

The sources, means of connection or separation may be within the unit.

NOTE 4: Where there is a potential hazard due to moving the unit whilst connected to an external installation, it is recommended that the unit is equipped with an electrical interlock, warning, alarm or other appropriate means to reduce the risk.

NOTE 5: For the purpose of this section, power inverters or frequency convertors supplied from the unit's electrical system or an auxiliary system driven by the unit's prime mover are also considered as generating sets.

Power inverters or frequency convertors shall include electrical separation where both the d.c. supply and the a.c. neutral point are earthed.

▼ **Figure 14.1** Example of connection to low voltage generating set located inside the unit, with or without an earth electrode

▼ **Figure 14.2** Example of connection to a low voltage generating set located outside the unit, with or without an earth electrode

▼ **Figure 14.3** Example of connection to a TN or TT electrical installation, with or without an earth electrode at the unit

NOTE: Where a PME earthing facility is used, see Regulation 717.411.4.

122 | Guidance Note 7: Special Locations
© The Institution of Engineering and Technology

▼ **Figure 14.4** Example of connection to a fixed electrical installation with any type of earthing system using a simple separation transformer and an internal IT system, with an earth electrode

▼ **Figure 14.5** Example of connection to a fixed electrical installation with any type of earthing system using a simple separation transformer and an internal TN system, with or without an earth electrode

▼ **Figure 14.6** Example of connection to a fixed electrical installation with any type of earthing system by using an IT system without automatic disconnection in the event of first fault

Key to Figures 14.1 to 14.6

1a Connection of the unit to a supply through a transformer with simple separation

1b Connection of the unit to a supply in which the protective measures are effective

1c Connection to an LV generator set in accordance with Section 551

2 Class II or equivalent enclosure up to the first protective device providing automatic disconnection of supply

4 Conductive external staircase, if any

5 Connection of the neutral point (or, if not available, a line conductor) to the conductive structure of the unit

6 Socket-outlets for use exclusively within the unit

6a Socket-outlets for use exclusively within the unit for reasons of continuity of supply in the event of first fault

6b Socket-outlets for general use if explicitly required (operation of the RCD in the event of first fault cannot be excluded)

7 Main equipotential bonding in accordance with Regulation 717.411.3.1.2

7a to an antenna pole, if any

7b to the conductive external stairs, if any, in contact with the ground

7c to a functional earth electrode, if required, depending on the functional requirements of the van and the design of the van and the location/event requirements

7d to the conductive structure of the unit

7e to an earth electrode for protective purposes, if required, depending on the functional requirements of the van and the design of the van and the location/event requirements

8 Protective devices, if required, for overcurrent and/or for protection by disconnection of supply in case of a second fault

9   Protective devices for overcurrent and for automatic disconnection of supply in case of a second fault

10   Socket-outlets for current-using equipment for use outside the unit

13   Current-using equipment for use exclusively within the unit

14   Overcurrent protective device, if required

15   Overcurrent protective device

16a   RCD having the characteristics specified in Regulation 415.1.1 for protection by automatic disconnection of supply for circuits of equipment for use outside the unit

16b   RCD for protection by automatic disconnection of supply for circuits of equipment for use inside the unit

18   Main earthing terminal or bar

21   Transformer with at least simple separation e.g. 230 V current-using equipment outside the unit

25   Insulation monitoring device or insulation fault location system including monitoring of the N conductor if distributed (disconnection only in the event of second fault)

28   Possible connection point to an existing lightning protection system in the vicinity (if any) for protection against lightning electromagnetic pulse (if any)

717.411.6.2   ### 14.4.2 IT system

An IT system can be provided by:

a   an isolating transformer or a low voltage generating set, with an insulation monitoring device or an insulation fault location system, both without automatic disconnection of the supply in case of the first fault and without a need of connection to an earthing installation (see Figure 14.6); the second fault shall be automatically disconnected by overcurrent protective devices according to
411.6.4   Regulation 411.6.4, or

b   a transformer providing simple separation, e.g. in accordance with BS EN 61558-1, with an RCD and an earth electrode installed to provide automatic disconnection in the case of failure in the transformer providing the simple separation (see Figure 14.4).

14.5 Protection against electric shock

717.410.3.5   The protective measures of obstacles and placing out of reach (Section 417) are not permitted.

717.410.3.6   The protective measure of non-conducting location (Regulation 418.1) is not permitted and earth-free local equipotential bonding (Regulation 418.2) is not recommended.

717.415.1   Every socket-outlet intended to supply current-using equipment outside the unit shall
415.1.1   be provided with additional protection by an RCD with a rated residual operating current not exceeding 30 mA, with the exception of socket-outlets which are supplied from circuits with protection by:

i   SELV, or

ii   PELV, or

iii electrical separation.

14.6 Protective equipotential bonding

717.411.3.1.2 Accessible conductive parts of the unit, such as the chassis, shall be connected through the main protective bonding conductors to the main earthing terminal within the unit.

The protective bonding conductors shall be finely stranded. Cable types H05V-K and H07V-K to BS 6004 are considered appropriate.

14.7 Wiring systems external to the mobile or transportable unit

717.52.1 Flexible cables in accordance with H07RN-F (BS 7919), or cables of equivalent design, having a minimum cross-sectional area of 2.5 mm^2 copper shall be used for connecting the unit to the supply.

The flexible cable shall enter the unit by an insulating inlet in such a way as to minimise the possibility of any insulation damage or fault which might energize the exposed-conductive-parts of the unit.

14.8 Wiring systems inside the mobile or transportable unit

717.52.2 The wiring system shall be installed using one or more of the following:

i Unsheathed flexible cable with thermoplastic or thermosetting insulation to BS 6004 or BS 7211 installed in conduit in accordance with the appropriate part of the BS EN 61386 series or in trunking or ducting in accordance with the appropriate part of the BS EN 50085 series
ii Sheathed flexible cable with thermoplastic or thermosetting insulation to BS 6004, BS 7211 or BS 7919, if precautionary measures are taken such that no mechanical damage is likely to occur due to any sharp-edged parts or abrasion.

14.9 Proximity to non-electrical services

717.528.3.4 No electrical equipment, including wiring systems, except ELV equipment for gas supply control, shall be installed in any gas cylinder storage compartment.

ELV cables and electrical equipment may only be installed within the LPG cylinder compartment if the installation serves the operation of the gas cylinder (e.g. indication of empty gas cylinder), or is for use within the compartment.

Such electrical installations and components shall be constructed and installed so that they are not a potential source of ignition.

Where cables have to run through such a compartment, they shall be protected against mechanical damage by installation within a conduit system complying with the appropriate part of the BS EN 61386 series or within a ducting system complying with the appropriate part of the BS EN 50085 series.

Appx 5 Where installed, this conduit or ducting system shall be able to withstand an impact equivalent to AG3 (high severity) without visible physical damage.

717.514 ## 14.10 Identification and notices

A permanent notice of durable material as to be likely to remain easily legible throughout the life of the installation shall be fixed to the unit in a prominent position, preferably adjacent to the supply inlet connector. The notice should state in clear and unambiguous terms the following information:

- ▶ The types of supply which may be connected to the unit
- ▶ The voltage rating of the unit
- ▶ The number of supplies, phases and their configuration
- ▶ The on-board earthing arrangement
- ▶ The maximum power requirement of the unit.

14.11 Other equipment

717.55.1 Connecting devices mounted, accessed or used outside the unit and used to connect the unit to the supply, or supply other equipment, shall comply with BS EN 60309-2 and shall meet the following requirements:

i Connecting devices shall be within an enclosure of insulating material

ii Connecting devices shall afford a degree of protection of not less than IP44

iii Enclosures containing the connecting devices shall provide a degree of protection of at least IP55 when no cable connections are made to the unit. When cable connections are made to the unit the enclosure shall provide a degree of protection not less than IP44.

717.55.3 Generating sets able to produce voltages other than SELV or PELV, mounted in a mobile unit, shall automatically be switched off in case of an accident to the unit (e.g. event causing the release of airbags). If this requirement is difficult to implement an emergency switch, easily accessible, shall be installed.

14.12 Routine maintenance and testing

The service duty of mobile and transportable units will vary with the type of unit and type of use (e.g. owner-operator or hire) but it is likely that frequent connecting and disconnecting combined with transporting will amount to the equivalent of rough service life. Frequency of use should therefore be an important factor in determining inspection and testing intervals. It is recommended that a visual inspection is carried out on the connecting cable and all plugs and socket-outlets before each and every transported use of the unit. The results of the visual inspection should be entered in a logbook as a permanent record of the condition of the electrical equipment. No repairs or extensions are acceptable on the external cable system and this should be replaced in its entirety if there are signs of damage or wear-and-tear (abrasion).

As a minimum, the unit electrical system should be inspected and tested annually, a report obtained on its condition and the necessary maintenance work, if any, implemented before the unit is next used. The recommendations of Part 6 of BS 7671 should be followed in this regard, together with the specific guidance given in

GN3 Guidance Note 3. If the unit duty is considered to be arduous, the inspection intervals should be reduced to cater for the particular conditions experienced.

All RCDs should be tested regularly by operating the test button and periodically by a proprietary test instrument to ensure they conform to the parameters of the relevant British Standard.

Appx 6 All tests should be tabulated for record purposes. The necessary forms required by Part 6 of BS 7671 must be provided to the person ordering the work, by the contractor or persons carrying out the inspection and tests.

Small-scale embedded generators (SSEG)

15

A wind generator. Courtesy Dr Sung

15.1 Introduction and the law

Persons carrying out electrical work in dwellings must comply with the Building Regulations 2000 (as amended) in England and Wales, in particular Part P (Electrical safety – dwellings).

The Building Regulations mentioned above do not apply in Scotland or Northern Ireland. In Scotland the requirements of the Building (Scotland) Regulations 2004 apply, in particular Regulation 9, and in Northern Ireland the Building Regulations (Northern Ireland) 2000 (as amended) apply.

Amendment 1 to the 17th Edition

There are no changes introduced by Amendment No 1 to the 17th Edition.

Introduction to the 17th Edition

Low voltage generating sets such as small-scale embedded generators are not classed as special locations or installations under BS 7671:2008. There are, however, regulations related to this type of installation in Section 551 'Low voltage generating sets'.

Sect 551

For domestic installations in England and Wales, small-scale embedded generators such as microCHP units are classified as special installations in Approved Document P (Electrical safety – dwellings) guidance on Part P of the Building Regulations 2000 (as amended).

15.1.1 The Electricity Safety, Quality and Continuity Regulations 2002

Within the lifetime of this edition of Guidance Note 7 the use of small-scale embedded generators (SSEG) is likely to become widespread. The Electricity Safety, Quality and Continuity Regulations 2002 exempt sources of energy with an electrical output not exceeding 16 A per phase at low voltage (230 V) from Regulations 22(1)(b) and 22(1)(d) of the Regulations, and this chapter covers such sources of energy.

In addition to the electrical output limit of 16 A per phase (sub-paragraph (a) of Regulation 22(2)), other requirements for such small generators are:

(b) the source of energy is configured to disconnect itself electrically from the parallel connection when the distributor's equipment disconnects the supply of electricity to the consumer's installation; and

(c) the person installing the source of energy ensures that the distributor is advised of the intention to use the source of energy in parallel with the network before, or at the time of, commissioning the source.

The requirements that still need to be met in Regulation 22(1) are:

… No person shall install or operate a source of energy which may be connected in parallel with a distributor's network unless he –

(a) has the necessary and appropriate equipment to prevent danger or interference with that network or with the supply to consumers so far as is reasonably practicable;

(c) where the source of energy is part of a low voltage consumer's installation, complies with the provisions of British Standard Requirements [meaning, by definition, BS 7671].

15.1.2 BS 7671:2008 Section 551 – Low voltage generating sets

551.2   Section 551 includes further requirements in Regulation 551.2 to ensure the safe connection of low voltage generating sets including small-scale embedded generators.

551.4.2   Regulation 551.4.2 requires the use of RCDs. This regulation states:

> The generating set shall be connected so that any provision within the installation for protection by RCDs in accordance with Chapter 41 remains *effective for every intended combination of sources of supply.*

551.1 Note   A note to Regulation 551.1 informs that the procedure for connecting generating sets up to 16 A in parallel with the public supply is given in ESQCR 2002. Requirements of the distributor for the connection of units rated up to 16 A are given in BS EN 50438 *Requirements for the connection of micro-generators in parallel with public low-voltage distribution networks.*

551.7.2   For sets above 16 A the requirements of the distributor must be ascertained. The 17th Edition recognises that there are two connection options:

     i   connection into a separate dedicated circuit
     ii   connection into an existing final circuit.

Connection into a dedicated circuit is preferred.

Regulation 551.7.2 sets out the requirements for the two options. The regulation requires that a generating set used as an additional source of supply in parallel with another source shall either be installed on the supply side of all protective devices for the final circuits of the installation (connection into a separate dedicated circuit) or, if connected on the load side of all protective devices for the final circuits, must fulfil a number of additional requirements. These additional requirements are:

i The current-carrying capacity of the final circuit conductors shall be greater than or equal to the rated current of the protective device plus the rated output of the generating set

ii A generating set shall not be connected to a final circuit by a plug and socket

iii A residual current device providing additional protection of the final circuit in accordance with Regulation 415.1.1 shall disconnect all live conductors including the neutral conductor

iv The line and neutral conductors of the final circuit and of the generating set shall not be connected to Earth

v Unless the device providing automatic disconnection of the final circuit in accordance with Regulation 411.3.2 disconnects the line and neutral conductors, it shall be verified that the combination of the disconnection time of the protective device for the final circuit and the time taken for the output voltage of the generating set to reduce to 50 V or less is not greater than the disconnection time required by Regulation 411.3.2 for the final circuit.

The generator equipment should be type-tested and approved by a recognised body.

15.2 Engineering Recommendation G83

To assist network operators and installers the Energy Networks Association has prepared Engineering Recommendation G83: *Recommendations for the connection of small-scale embedded generators (up to 16 A per phase at low voltage) in parallel with public low voltage distribution networks*. The guidance given in this chapter is intended to reproduce the requirements of the Engineering Recommendation as they would apply to persons responsible for electrically connecting such generators.

Engineering Recommendation G83 is for all small-scale embedded generator (SSEG) installations with an output up to and including 16 A per phase, single or multiphase, 230/400 V a.c. This includes:

▶ domestic combined heat and power
▶ hydro
▶ wind power
▶ photovoltaic
▶ fuel cells.

Requirements for solar photovoltaic (PV) installations are found in Chapter 10 of this Guidance Note (see also section 15.7.2).

The Engineering Recommendation describes two types of connection procedure for the connection of an SSEG unit(s) in parallel with the public low voltage distribution network.

The first is for a single SSEG unit referred to as a Stage 1 connection within a single customer's installation. The installer firstly has to make the distribution network operator (DNO) aware of the SSEG installation at or before the time of commissioning, in accordance with the ESQC Regulation 22(2)(c). The installer then has to provide the DNO with all the necessary information on the installation within 30 days of the SSEG unit being commissioned.

The second is for a planned installation project referred to as a Stage 2 connection where there is a proposal to install multiple SSEG units in a close geographical region. The installer would be expected to discuss the project with the local DNO prior to installation, who will assess the impact of the connections on the network and provide the appropriate connection conditions. The installer then has to provide the DNO with all the necessary information on the installation within 30 days of commissioning.

Engineering Recommendation G83 incorporates forms, which define the information required by a public distribution network operator for a small-scale embedded generator which is connected in parallel with a public low voltage distribution network. Supply of information in this form, for a suitably type-tested unit, is intended to satisfy the legal requirements of the DNO and hence will satisfy the legal requirements of the Electricity Safety, Quality and Continuity Regulations 2002.

15.3 Installation and wiring

The installation that connects the embedded generator to the supply terminals must comply with BS 7671.

A suitably rated overcurrent protective device shall protect the wiring between the electricity supply terminals and the embedded generator.

The SSEG must be connected directly to a local isolating switch. For single-phase machines the phase and neutral are to be isolated and for multiphase machines all phases and neutral are to be isolated. In all instances the switch, which must be manual, shall be capable of being secured in the 'off' isolating position. The switch is to be located in an accessible position in the customer's installation.

15.4 Means of isolation

551.7.6   Regulation 551.7.6 states the requirements for isolation:

> **551.7.6**   Means shall be provided to enable the generating set to be isolated from the system for distribution of electricity to the public. For a generating set with an output exceeding 16 A, the accessibility of this means of isolation shall comply with national rules and distribution system operator requirements. For a generating set with an output not exceeding 16 A, the accessibility of this means of isolation shall comply with BS EN 50438.

15.5 Earthing

When a SSEG is operating in parallel with a distributor's network, there shall be no direct connection between the generator winding (or pole of the primary energy source in the case of PV array or fuel cell) and the network operator's earth terminal. See Figure 15.1.

▼ Figure 15.1
Earthing of parallel
operation SSEG

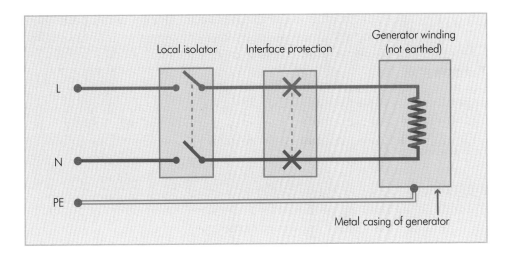

Chap 54   For all connections earthing arrangements shall comply with the requirements of BS 7671.

15.6   Labelling

514.15.1   Labels (i.e. warning notices) are required at:

▶ the supply terminals (fused cut-out),
▶ the meter position,
▶ the consumer unit or distribution board to which the SSEG is connected, and
▶ all points of isolation of all sources of supply to indicate the presence of the SSEG within the premises.

The Health and Safety (Safety Signs and Signals) Regulations 1996 stipulate that the labels should display the prescribed triangular shape and size using black on yellow colouring. A typical label both for size and content is shown in Figure 15.2.

▼ Figure 15.2
Warning notice for
alternative supplies

In addition, Engineering Recommendation G83 requires up-to-date information to be displayed at the point of connection with a distributor's network as follows:

a   A circuit diagram showing the circuit wiring, including all protective devices, between the embedded generator and the network operator's fused cut-out. This diagram is also required to show by whom the generator is owned and maintained

b   A summary of the protection settings incorporated within the equipment.

Figure 15.3 is an example of the type of circuit diagram that needs to be displayed. This diagram is for illustrative purposes and not intended to be fully descriptive. Metering devices, consumer unit or circuit protective devices within the installation would need to be shown.

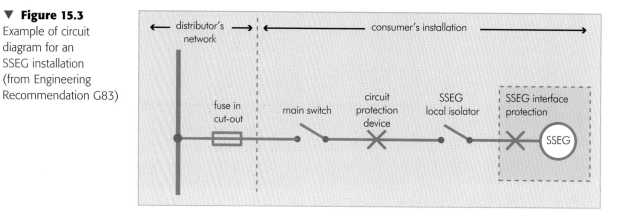

The installer is required to advise that it is the user's responsibility to ensure that this safety information is kept up to date. The installation operating instructions shall contain the SSEG manufacturer's contact details, e.g. name, telephone number and web address.

15.7 Types of SSEG

Note: SSEGs will only operate in parallel with the mains supply.

15.7.1 Domestic combined heat and power

Annex B of Engineering Recommendation G83 specifies the particular requirements for combined heating and power sets. These are likely to be the most common type of set encountered by the electricity installer. They can be incorporated into a household gas boiler to generate electricity.

Some small combined heat and power generators incorporate a Stirling engine – see Figure 15.4.

▼ **Figure 15.4**
The Stirling engine cycle

The Stirling engine does not burn the gas within the cylinder as in an internal combustion engine. The power to the engine is delivered by the combustion gases from the gas burner, and the energy transfer is in effect achieved by the temperature difference between the burner exhaust gases and burner input air or circulating water.

The basic principle is that the sealed gas within the engine is heated by the burner gases and expands. On expansion the driven piston generates electricity in the winding and compresses the planar spring. In the expanded, say, down position of the piston the gas within the piston is subject to the cooling effect of the cooler input air (or circulated water) and contracts, assisted by the planar spring. At the compression position the gas in the cylinder is now heated by the burner gases only and is not subject to the cooling effects of the input air.

15.7.2 Solar photovoltaic (PV) power supply systems

Chapter 10 of this publication gives guidance for solar photovoltaic power supply systems. The requirements are for systems that will generate only when run in parallel with the electricity supply. Requirements for photovoltaic power supply systems which are intended for stand-alone operation are under consideration by the IEC and not considered here.

Temporary electrical installations for structures, amusement devices and booths at fairgrounds, amusement parks and circuses

16

16.1 Introduction

Introduction to Amendment No 1 to the 17th Edition

Sect 740 Section 740 of Amendment 1 to BS 7671:2008 includes a number of notable changes.

Under Scope, reference is now made to BS 7909 *Code of practice for temporary electrical systems for entertainment and related purposes*. Under the protective measure

740.411.4 'Automatic disconnection of supply', TN system, Regulation 740.411.4 has been modified, and now refers to PME. The regulation also prohibits the use of a PEN conductor.

The changes introduced by the 17th Edition (BS 7671:2008) when it was published in January 2008 are mentioned below for information.

Introduction to the 17th Edition

This is a new section that has been introduced into the 17th Edition and it also appears in this Guidance Note for the first time.

Section 740 recognises that some installations are exposed to many differing and onerous circumstances, as they are often frequently installed, dismantled, moved to a new location then installed and operated again.

To compound problems, such installations can be exposed to the elements, and the locations are open to the general public, may house animals and livestock, and are also a place of work.

The equipment must function without compromising safety; therefore, the installation has to be fit for purpose and be designed to cope with ever-changing conditions.

Section 740 prescribes particular measures to reduce the risks caused by this harsh treatment of the electrical installation.

16.2 Scope

740.1.1 Section 740 specifies the minimum electrical installation requirements to facilitate the safe design, installation and operation of temporarily erected mobile or transportable electrical machines and structures which incorporate electrical equipment. The machines and structures are intended to be installed repeatedly, without loss of safety, temporarily, at fairgrounds, amusement parks, circuses or similar places.

Note 1: The permanent electrical installation, if there is one, is excluded from the scope.

The object of this section is to define the electrical installation requirements for such structures and machines, being either integral parts or constituting the total amusement device. The scope does not include the internal electrical wiring of machines (see BS EN 60204-1).

Note 2: Guidance on temporary electrical installations for entertainment related purposes is given in BS 7909.

16.3 Electrical supplies

740.313.1.1 The nominal supply voltage of temporary electrical installations in booths, stands and amusement devices should not exceed 230/400 V a.c. or 440 V d.c. Supplies can be obtained from a number of sources:

i from the public network, i.e. the DNO
ii generators, i.e. those mounted on trucks owned by the touring event
iii from privately owned supplies, i.e. a local factory with sufficient spare capacity.

740.313.3 There can be any number of electrical sources supplying the temporary system and it is of paramount importance that line and neutral conductors from different sources are not interconnected.

Where the supply is obtained from the DNO any instructions given must be adhered to. Supplies obtained from the DNO would preferably be TN-S but this isn't always possible. A TN-S system has the neutral of the source of energy connected with Earth at one point only, at, or as near as is reasonably practicable to, the source of supply.

The consumer's main earthing terminal is typically connected to the metallic sheath of the distributor's SWA service cable.

740.411.4.1 A PME earthing facility shall not be used as the means of earthing for an installation falling within the scope of this section.

> **Note:** The ESQCR prohibit the use of a PME earthing facility as the means of earthing for the installation of a caravan or similar construction.

740.411.4.3 Where the type of system earthing is TN, a PEN conductor shall not be used downstream of the origin of the temporary electrical installation.

740.411.6 Where continuity of service is important, IT systems may be used for d.c. applications only.

16.4 Protection against electric shock

740.410.3 At the origin of each electrical supply, to all or part of the installation, an RCD, with a rated residual operating current not exceeding 300 mA, is to be installed to provide automatic disconnection of supply. As there will be further RCDs downstream of this point, this RCD should be of the S-type complying with the requirements of BS EN 61008-1 or BS EN 61009-1, and incorporate a time delay in accordance with BS EN 60947-2, to provide discrimination with further RCDs protecting final circuits. For supplies to a.c. motors, RCDs should be the time-delayed or S-type where necessary to prevent unwanted tripping.

740.410.3.5 The protective measure of obstacles is not permitted in this type of installation; however, placing out of arm's reach is acceptable for electric dodgems – see section 16.8.2.

740.410.3.6 The protective measures of non-conducting location (Regulation 418.1) and earth-free local equipotential bonding (Regulation 418.2) are not permitted.

16.5 Additional protection

740.415.1 All final circuits in the installation for lighting, socket-outlets rated up to 32 A, and mobile equipment connected by a flexible cable and rated up to 32 A, are to be protected by an RCD having a rated residual operating current not exceeding 30 mA.

The requirement for additional protection relates to the increased risk of damage to cables within an installation of this nature.

Lighting circuits incorporating emergency luminaires, with self-contained batteries for example, should be protected by the same RCD protecting that lighting circuit.

Additional protection is not required for:

i SELV or PELV circuits – this measure alone is deemed to be a protective measure in all situations

ii circuits protected by electrical separation

iii lighting circuits placed out of arm's reach – provided they are not supplied by socket-outlets for household or similar purposes, e.g. those manufactured to BS 1363, or socket-outlets according to BS EN 60309-1. Typically, this would be achieved by devices for connection of luminaires to BS EN 61995 or luminaire supporting couplers to BS 6972.

740.415.2 To minimise potentials, supplementary bonding should be installed to connect all exposed-conductive-parts and extraneous-conductive-parts that can be touched by livestock. Where a metal grid is laid in the floor, or extraneous-conductive-parts are accessible, they should be included within the supplementary bonding of the location. It is important to note that animal excrement and urine are very corrosive and so all supplementary bonding connections should be protected by suitable enclosure.

16.6 Installation

16.6.1 Wiring systems

740.521.1 Conduit, cable trunking, cable ducting, tray and ladder systems should comply with the following standards:

  i   Conduit systems – BS EN 61386 series
  ii Cable trunking and ducting systems – the relevant part 2 of BS EN 50085
  iii Tray and ladder systems – BS EN 61537.

Buried cables should be protected against mechanical damage. Conduit classified as 450N regarding protection against compression and classified as normal regarding protection against impact, according to BS EN 50086-2-4, would fulfil this requirement.

Mechanical protection should be used in all public areas and in areas where the wiring system crosses roads or walkways. Where mechanical protection is provided:

  i conduit systems should comply with BS EN 61386-21 with a classification of heavy protection against compression and heavy protection against impact. Metallic and composite conduit systems should be class 3 regarding protection against corrosion, e.g. medium protection inside and high protection outside.
  ii Cable trunking and cable ducting systems should comply with BS EN 50085 series with a classification 5J regarding protection against impact.

Where subject to movement, the wiring system should be of flexible construction and should comply with BS EN 61386-23. This should use cables of type H07RNF or H07BN4-F (BS7919).

16.6.2 Cables

740.521.1 All cables should be fire rated and meet the requirements of BS EN 60332-1-2. Cables of type H07RNF or H07BN4-F (BS 7919) together with conduit complying with BS EN 61386-23 are deemed to satisfy this requirement.

Cables should have a minimum rated voltage of 450/750 V, except that, within amusement devices, cables having a minimum rated voltage of 300/500 V may be used.

Where cables are buried in the ground, the route should be marked at suitable intervals and be protected against mechanical damage. Where there is a risk of mechanical damage, armoured cables or cables protected against mechanical damage due to external influence of medium severity or greater (e.g. > AG2) should be used.

16.6.3 Electrical connections

740.526 Joints should not be made in cables except where necessary as a connection into a circuit. Where joints are made, these should either use connectors in accordance with the relevant British Standards and the manufacturer's instructions or the connection should be made in an enclosure with a degree of protection of at least IPXXD or IP4X. Where strain can be transmitted to terminals the connection should incorporate cable anchorage(s).

16.6.4 External influences

740.512.2   Electrical equipment should have a degree of protection of at least IP44.

16.6.5 Switchgear and controlgear

740.51   Switchgear and controlgear should be placed in cabinets which can be opened only by the use of a key or a tool, except for those parts designed and intended to be operated

Part 2   by ordinary persons.

16.6.6 Isolation

740.537.1   It is a requirement that every electrical installation of a booth, stand or amusement device has its own means of isolation, switching and overcurrent protection; these devices should be readily accessible.

There are similar requirements for supplies to amusement devices.

Additionally, each distribution circuit should be provided with its own readily accessible and properly identified means of isolation. A device for isolation should disconnect all live conductors – line(s) and neutral conductors. Examples of devices that can be used for isolation where marked as suitable are:

i   circuit-breakers
ii   RCDs
iii   plug and socket arrangements.

Table 53.4   **Note:** Table 53.4 of BS 7671 includes guidance on device types that are suitable for isolation.

16.6.7 Luminaires

740.55.1.1   Every luminaire and decorative lighting chain should have a suitable IP rating and be securely attached to the structure or support intended to carry it. Its weight should not be carried by the supply cable, unless it has been selected and erected for this purpose. Luminaires and decorative lighting chains mounted less than 2.5 m, i.e. within arm's reach, above floor level or which could be otherwise accessible to incidental contact, should be firmly fixed, sited and guarded to prevent risk of injury to persons or ignition of materials.

Access to the fixed light source should only be possible after removing a barrier or an enclosure, which should only be possible by the use of a tool.

Lighting chains should use H05RN-F (BS 7919) cable or equivalent, and may be used in any length provided the overcurrent protective device in the circuit is correctly rated.

740.55.1.2   Insulation-piercing lampholders must not be used unless the cables and lampholders are compatible and the lampholders are non-removable once fitted to the cable.

740.55.3   Luminous tubes, signs or lamps with an operating voltage higher than 230 V/400 V a.c., e.g. neon signs, are to be installed out of arm's reach or be adequately protected from accidental or deliberate damage, to reduce the risk of injury to persons.

740.55.1.3 A separate circuit should be used which should be controlled by an emergency switch. The switch should be easily visible, accessible and marked in accordance with the requirements of the local authority. Lamps in shooting galleries and other sideshows where projectiles are used should be suitably protected against accidental damage.

740.55.1.4 Where transportable floodlights are used, they should be mounted so that the luminaire is inaccessible to non-instructed persons. Supply cables should be flexible and have adequate protection against mechanical damage.

16.6.8 Safety isolating transformers and electronic convertors

740.55.5 Safety isolating transformers should comply with BS EN 61558-2-6 or provide an equivalent degree of safety.

Each transformer or electronic convertor should incorporate a protective device which can be manually reset only; this device should protect the secondary circuit.

Safety isolating transformers should be mounted out of arm's reach or be mounted in a location that provides equal protection, e.g. in a panel or room that can only be accessed by a skilled or instructed person, and should have adequate ventilation.

Access by competent persons for testing or by a skilled person competent in such work for protective device maintenance should be provided. Electronic convertors should conform to BS EN 61347-2-2.

Enclosures containing rectifiers and transformers should be adequately ventilated and the vents should not be obstructed when in use.

16.6.9 Plugs and socket-outlets

740.55.7 An adequate number of socket-outlets should be installed to allow user requirements to be met safely.

In booths, stands and for fixed installations, one socket-outlet for each square metre or linear metre of wall is generally considered adequate.

Socket-outlets dedicated to lighting circuits placed out of arm's reach should be labelled according to their purpose.

When used outdoors, plugs, socket-outlets and couplers should comply with BS EN 60309-2, or, where interchangeability is not required, BS EN 60309-1.

16.7 Fire risk

GN4 Additional guidance for protection against fire for this type of location can be found in section 8.4 of Guidance Note 4: *Protection Against Fire*.

16.7.1 Luminaires and floodlights

740.55.1.5 Luminaires and floodlights should be so fixed and protected that a focusing or concentration of heat is not likely to cause ignition of any material.

16.7.2 Electric motors

422.3.7 An electric motor which is automatically or remotely controlled or which is not continuously supervised should be fitted with a manual reset protective device against excess temperature.

16.8 Electrical equipment

16.8.1 Electrical supply to devices

740.55.8 At each amusement device, there should be a connection point readily accessible and permanently marked to indicate the following essential characteristics:

  i   Rated voltage
  ii Rated current
  iii Rated frequency.

16.8.2 Electric dodgems

740.55.9 Electric dodgems should only be operated at voltages not exceeding 50 V a.c. or 120 V d.c. The circuit should have electrical separation from the electrical supply by means of a safety isolating transformer in accordance with BS EN 61558-2-4 or a motor-generator set.

16.8.3 Low voltage generating sets

740.551 It is very important that all generators are located or protected so as to prevent danger and injury to people through inadvertent contact with hot surfaces and dangerous parts.

The electrical equipment associated with the generator should be mounted securely and, if necessary, on anti-vibration mountings.

542.2.4 Where a generator supplies a temporary installation, forming part of a TN, TT or IT system, care should be taken to ensure that the earthing arrangements are adequate and, in cases where earth electrodes are used, they are considered to be continuously effective. In reality, this means that the drying of the ground in summer, or freezing of the ground in winter, should not adversely affect the value of earth fault loop impedance for the installation. The neutral conductor of the star-point of the generator should, except for IT systems, be connected to the exposed-conductive-parts of the generator.

16.9 Inspection and testing

16.9.1 The temporary installation

740.6 The electrical installation between its origin and any electrical equipment should be inspected and tested after each assembly on site.

Internal electrical wiring of roller coasters, electric dodgems, etc., is not considered as part of the verification of the electrical installation. In special cases the number of the tests may be modified according to the type of temporary electrical installation.

16.9.2 Amusement devices

The scope of Section 740 does not cover amusement devices, but, in law, there is still a requirement to ensure that the devices are fit for use.

The Amusement Device Safety Council (ADSC) is the policy-making body for safety, self-regulation and technical guidance in the UK amusement industry. This committee, in partnership with HSE, develops the policy for and oversees the Amusement Device Inspection Procedures Scheme (ADIPS). ADIPS is the fairground and amusement park industry's self-regulated safety inspection scheme which registers competent ride inspectors and the rides they inspect. The purpose of the scheme is to promote and improve fairground and amusement park safety through rules and procedures relating to the annual inspection of the amusement devices. It is supported by industry associations who require that their members use ADIPS. ADIPS is also available to any operator of amusement devices as it demonstrates 'best practice' and their compliance with the Health and Safety at Work etc. Act 1974.

Operating and maintenance gangways

17

17.1 Introduction

Sect 729 This is a new section that has been introduced by Amendment No 1 to the 17th Edition and it also appears in this Guidance Note for the first time.

132.12 Section 729 applies to restricted areas. These are areas such as switchrooms with switchgear and controlgear assemblies with a need for operating or maintenance gangways for authorized persons. The fundamental requirement for accessibility of electrical equipment in BS 7671:2008 is contained in Chapter 13. Regulation 132.12 states: Electrical equipment shall be arranged so as to afford as may be necessary: (i) sufficient space for the initial installation and later replacement of individual items of electrical equipment (ii) accessibility for operation, inspection, testing, fault detection, maintenance and repair.

Regulation 15 of the Electricity at Work Regulations 1989 has requirements for working space, access and lighting and requires that, for the purposes of enabling injury to be prevented, adequate working space, adequate means of access and adequate lighting shall be provided at all electrical equipment on which or near which work is being done in circumstances which may give rise to danger (Regulation 15 applies whilst there is work activity ongoing). Regulation 14 of the Electricity at Work Regulations is concerned with work on or near any live conductors.

Sect 417 See also Section 417 of BS 7671:2008.

Section 729 prescribes particular measures to reduce the risks in these restricted areas.

17.2 Scope

The particular requirements of this section apply to basic protection and other aspects relating to the operation or maintenance of switchgear and controlgear within restricted areas including gangways.

17.3 Accessibility
729.513

17.3.1 Requirements for operating and maintenance gangways

729.513.2 Regulation 729.513.2 requires that the width of gangways and access areas shall be adequate for work, operational access, emergency access, emergency evacuation and for the movement of equipment.

17.3.2 Restricted access areas where basic protection is provided by barriers or enclosures

729.513.2.1 In restricted access areas where basic protection is provided by barriers or enclosures Regulation 729.513.2.1 gives the following minimum distances (see Fig 17.1):

    i   Gangway width of 700 mm between: barriers or enclosures and switch handles or circuit-breakers in the most onerous position, and barriers or enclosures or switch handles or circuit-breakers in the most onerous position and the wall

    ii Gangway width of 700 mm between barriers or enclosures or other barriers or enclosures and the wall

    iii Height of gangway to barrier or enclosure above floor (minimum dimension 2000 mm)

417.3   **iv** Live parts placed out of reach, see Regulation 417.3 (minimum dimension 2500 mm).

Note: Where additional workspace is needed e.g. for special switchgear and controlgear assemblies, larger dimensions may be required.

17.3.3 Restricted access areas where the protective measure of obstacles is applied

729.513.2.2   In restricted access areas where the protective measure of obstacles applies Regulation 729.513.2.2 gives the following minimum distances (see Fig 17.2):

    i   Gangway width of 700 mm between: obstacles and switch handles or circuit-breakers in the most onerous position, and obstacles or switch handles or circuit-breakers in the most onerous position and the wall.

    ii Gangway width of 700 mm between obstacles or other obstacles and the wall

    iii Height of gangway to obstacles, above floor (minimum dimension 2000 mm)

417.3   **iv** Live parts placed out of reach, see Regulation 417.3 (minimum dimension 2500 mm)

Note: The HSE would not expect to see any new switchboards installed in the UK where there was potential access to live exposed busbars.

17.3.4 Access of gangways

729.513.2.3   Regulation 729.513.2.3 has requirements for access of gangways. For closed restricted access areas with a length exceeding 6 m, accessibility from both ends is recommended. However, gangways longer than 10 m must be accessible from both ends.

The regulation recognises that this may be accomplished by placement of the equipment a minimum of 700 mm from all walls or by providing an access door, if needed, on the wall against which the equipment is positioned. However, closed restricted access areas with a length exceeding 20 m must be accessible by doors from both ends.

729 annex A   Finally, Annex A729 (note: normative, i.e. obligatory) contains a number of additional requirements for closed restricted access areas in order to permit easy evacuation. The annex considers two cases; these include:

▶ a minimum passing width of 700 mm with circuit-breakers in the isolated position and all doors closed, and

▶ a minimum passing width of 500 mm with circuit-breakers fully withdrawn and doors fully open.

Persons involved in this work are advised to seek advice from the HSE.

▼ **Figure 17.1** Gangways in installations with protection by barriers or enclosures

Note: The above dimensions apply after barriers and enclosures have been fixed and with circuit-breakers and switch handles in the most onerous position, including "isolation".

▼ **Figure 17.2** Gangways in installations with protection by obstacles

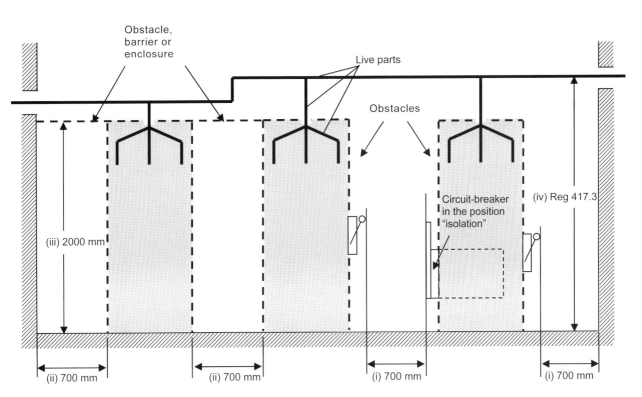

Note: The above dimensions apply after all obstacles, barriers and enclosures have been fixed and with circuit-breakers and switch handles in the most onerous position, including "isolation".

Index

IET Wiring Regulations and associated publications

The IET prepares regulations for the safety of electrical installations, the IET Wiring Regulations (BS 7671 *Requirements for Electrical Installations*), which are the standard for the UK and many other countries. The IET also offers guidance around BS 7671 in the form of the Guidance Notes series and the Electrician's Guides as well as running a technical helpline. The Wiring Regulations and guidance are now also available as e-books through Wiring Regulations Digital (see overleaf).

IET Members receive discounts across IET publications and e-book packages.

Requirements for Electrical Installations, IET Wiring Regulations 17th Edition (BS 7671:2008 incorporating Amendment No. 1:2011)
Order book PWR1701B Paperback 2011
ISBN: 978-1-84919-269-9 **£80**

On-Site Guide BS 7671:2008 (2011)
Order book PWGO171B Paperback 2011
ISBN: 978-1-84919-287-3 **£24**

IET Guidance Notes

The IET also publishes a series of Guidance Notes enlarging upon and amplifying the particular requirements of a part of the IET Wiring Regulations.

Guidance Note 1: Selection & Erection, 6th Edition
Order book PWG1171B Paperback 2012
ISBN: 978-1-84919-271-2 **£32**

Guidance Note 2: Isolation & Switching, 6th Edition
Order book PWG2171B Paperback 2012
ISBN: 978-1-84919-273-6 **£27**

Guidance Note 3: Inspection & Testing, 6th Edition
Order book PWG3171B Paperback 2012
ISBN: 978-1-84919-275-0 **£27**

Guidance Note 4: Protection Against Fire, 6th Edition
Order book PWG4171B Paperback 2012
ISBN: 978-1-84919-277-4 **£27**

Guidance Note 5: Protection Against Electric Shock, 6th Edition
Order book PWG5171B Paperback 2012
ISBN: 978-1-84919-279-8 **£27**

Guidance Note 6: Protection Against Overcurrent, 6th Edition
Order book PWG6171B Paperback 2012
ISBN: 978-1-84919-281-1 **£27**

Guidance Note 7: Special Locations, 4th Edition
Order book PWG7171B Paperback 2012
ISBN: 978-1-84919-283-5 **£27**

Guidance Note 8: Earthing & Bonding, 2nd Edition
Order book PWG8171B Paperback 2012
ISBN: 978-1-84919-285-9 **£27**

Electrician's Guides

Electrician's Guide to the Building Regulations Part P, 2nd Edition
Order book PWGP170B Paperback 2008
ISBN: 978-0-86341-862-4 **£22**

continues overleaf ▶

Electrical Installation Design Guide
Order book PWR05030 Paperback 2008
ISBN: 978-0-86341-550-0 **£22**
To be updated 2012

Electrician's Guide to Emergency Lighting
Order book PWR05020 Paperback 2009
ISBN: 978-0-86341-551-7 **£22**

Electrician's Guide to Fire Detection and Alarm Systems
Order book PWR05130 Paperback 2010
ISBN: 978-1-84919-130-2 **£22**

Other guidance

Commentary on IEE Wiring Regulations (17th Edition, BS 7671:2008)
Order book PWR08640 Hardback 2009
ISBN: 978-0-86341-966-9 **£65**

Electrical Maintenance, 2nd Edition
Order book PWR05100 Paperback 2006
ISBN: 978-0-86341-563-0 **£40**

Code of Practice for In-service Inspection and Testing of Electrical Equipment, 3rd Edition
Order book PWR08630 Paperback 2007
ISBN: 978-0-86341-833-4 **£40**

Electrical Craft Principles, Volume 1, 5th Edition
Order book PBNS0330 Paperback 2009
ISBN: 978-0-86341-932-4 **£25**

Electrical Craft Principles, Volume 2, 5th Edition
Order book PBNS0340 Paperback 2009
ISBN: 978-0-86341-933-1 **£25**

For more information and to buy the IET Wiring Regulations and associated guidance, visit www.theiet.org/electrical

Wiring Regulations Digital

The IET Wiring Regulations and associated guidance are now available in e-book format. You can search and link within and between books, make your own notes and print pages from the books. There are two packages available:

Wiring Regulations Digital: Domestic
BS 7671:2008 (2011)*, the On-Site Guide, Guidance Notes 1 & 3 and the Electrician's Guide to the Building Regulations. **£175 + VAT**

Wiring Regulations Digital: Industrial
BS 7671:2008 (2011)*, the On-Site Guide, Guidance Notes 1–8, the Electrician's Guide to the Building Regulations and the Electrical Installation Design Guide. **£350 + VAT**

Find out more and order at www.theiet.org/digital-regs
* You will initially receive the 17th Edition version of each of these e-books and any published Amendment No. 1 titles to date. Your package will be updated for free to each new Amendment No. 1 title upon publication.

Electrical **excellence**

Order Form

How to order

BY PHONE:
+44 (0)1438 767328
BY FAX:
+44 (0)1438 767375
BY POST:
The Institution of
Engineering
and Technology,
PO Box 96,
Stevenage
SG1 2SD, UK
ONLINE:
www.theiet.org/electrical

INFORMATION SECURITY
Please do not submit your form by
email. The IET takes the security of your
personal details and credit/debit card
information very seriously and will not
process email transactions.

Free postage within the UK. Outside
UK (Europe) add £5.00 for first title
and £2.00 for each additional book.
Rest of World add £7.50 for the first
book and £2.00 for each additional
book. Books will be sent via airmail.
Courier rates are available on request,
please call +44 (0) 1438 767328
or email sales@theiet.org for rates.

**Member Discounts: These cannot be
used in conjunction with any other IET
discount offers.

GUARANTEED RIGHT OF RETURN:
If at all unsatisfied, you may return
books in new condition within 30 days
for a full refund. Please include a copy
of the invoice.

DATA PROTECTION:
The information that you provide to the
IET will be used to ensure we provide
you with products and services that
best meet your needs. This may include
the promotion of specific IET products
and services by post and/or electronic
means. By providing us with your email
address and/or mobile telephone
number you agree that we may contact
you by electronic means. You can
change this preference at any time
by visiting www.theiet.org/my.

Details

Name:

Job Title:

Company/Institution:

Address:

Postcode: Country:

Tel: Fax:

Email:

Membership No (if Institution member):

Ordering information

Quantity	Book No.	Title/Author	Price (£)
		Subtotal	
		Member discount **	
		+ Postage /Handling*	
		Total	

Payment methods

☐ By **cheque** made payable to The Institution of Engineering and Technology

☐ By **credit/debit card:** ☐☐☐☐ ☐☐☐☐ ☐☐☐☐ ☐☐☐☐

☐ Visa ☐ Mastercard ☐ American Express ☐ Maestro Issue No: _____

Valid from: ☐☐ ☐☐   Expiry Date: ☐☐ ☐☐   Card Security Code: ☐☐☐☐
(3 or 4 digits on reverse of card)

Signature _____ Date_____

Cardholder Name:

Cardholder Address:

Town: Postcode:

Country:

Tel:

Email:

The Institution of Engineering and Technology is registered as a Charity in England & Wales (no 211014) and Scotland (no SC038698). Michael Faraday House, Six Hills Way, Stevenage, SG1 2AY

Dedicated website

Everything that you need from the IET is now in one place. Ensure that you are up-to-date with BS 7671 and find guidance by visiting our dedicated website for the electrical industry.

Catch up with the latest forum discussions, download the Wiring Regulations forms (as listed in BS 7671) and read articles from the IET's **free** Wiring Matters magazine.

The IET Wiring Regulations BS 7671:2008 (2011) and all associated guidance publications can be bought directly from the site. You can even pre-order titles that have not yet been published at discounted prices. **www.theiet.org/electrical**

Membership

Passionate about your career? Become an IET Member and benefit from a range of benefits from the industry experts. As co-publishers of the IET Wiring Regulations, we can assist you in demonstrating your technical professional competence and support you with all your training and career development needs.

The Institution of Engineering and Technology is the professional home for life for engineers and technicians. With over 150,000 members in 127 countries, the IET is the largest professional body of engineers in Europe.

Joining the IET and having access to tailored products and services will become invaluable for your career and can be your first step towards professional qualifications.

You can take advantage of...

- a 35% discount on BS 7671:2008 (2011) IET Wiring Regulations, associated guidance publications and Wiring Regulations Digital
- career support services to assist throughout your professional development
- dedicated training courses, seminars and events covering a wide range of subjects and skills
- an array of specialist online communities
- professional development events covering a wide range of topics from essential business skills to time management, health and safety, life skills and many more
- access to over 100 Local Networks around the world
- live IET.tv event footage
- instant on-line access to over 70,000 books, 3,000 periodicals and full-text collections of electronic articles through the Virtual library, wherever you are in the world.

Join online today: **www.theiet.org/join** or contact our membership and customer service centre on +44 (0)1438 765678.